GRADES 6-8

Mathematics
Assessment Sampler

Mathematics Assessment Samplers

A series edited by Anne M. Collins

GRADES 6-8

Mathematics Assessment Sampler

Items Aligned with NCTM's
Principles and Standards for School Mathematics

John Burrill, *Editor*
Hales Corners, Wisconsin

Anne M. Collins, *Series Editor*
Boston College Mathematics Institute, Chestnut Hill, Massachusetts

WRITING TEAM
Beth Cole
James Hirstein
Elizabeth Jones
Marge Petit

CONTRIBUTORS
Marthe Craig
Mary Eich
Carl Lager

NATIONAL COUNCIL OF
TEACHERS OF MATHEMATICS

Copyright © 2005 by
THE NATIONAL COUNCIL OF TEACHERS OF MATHEMATICS, INC.
1906 Association Drive, Reston, VA 20191-1502
(703) 620-9840; (800) 235-7566; www.nctm.org
All rights reserved

Library of Congress Cataloging-in-Publication Data

Mathematics assessment sampler, grades 6–8 : items aligned with NCTM's principles and standards for school mathematics / John Burrill, editor ; writing team, Beth Cole ... [et al.].
    p. cm. — (Mathematics assessment samplers)
  Includes bibliographical references.
  ISBN 0-87353-580-4
1. Mathematics—Study and teaching (Middle school) 2. Mathematical ability—Testing. I. Burrill, John (John C.) II. Series.
  QA11.2.M2776 2005
  510'.71'2—dc22
                                        2005013337

The National Council of Teachers of Mathematics is a public voice of mathematics education, providing vision, leadership, and professional development to support teachers in ensuring mathematics learning of the highest quality for all students.

PRINTED IN THE UNITED STATES OF AMERICA

Contents

On the Cover vi

Preface vii

Acknowledgments ix

About This Series xi

Introduction: About This Book 1

Chapter 1: Number and Operations 3

Chapter 2: Algebra 45

Chapter 3: Geometry 109

Chapter 4: Measurement 141

Chapter 5: Data Analysis and Probability 187

Chapter 6: Professional Development 223

Appendix: Items Matrices 243

Bibliography 257

Sources for Assessment Items 261

On the Cover

The term *sampler* comes from the Latin *exemplum*, meaning "an example to be followed, a pattern, a model or example." The earliest known samplers date to the sixteenth century, although samplers were probably stitched long before that time. Beginning in the mid–eighteenth century, young girls commonly worked samplers as part of their education. During Victorian times, samplers metamorphosed into decorative articles hung by proud parents on parlor walls. As designs became more elaborate, generally only one stitch remained in use, leading to the well–known cross–stitch samplers of today.

The electronically "stitched" sampler on the cover highlights the relationships among knowledge (owl), learning (school), and the NCTM Standards (logo). The number patterns embedded in the border design (the counting sequence across the top; the Fibonacci sequence 1, 1, 2, 3, 5, 8, ... around the left, bottom, and right borders) echo pattern motifs seen in samplers in earlier times.

The Mathematics Assessment Samplers are intended to give teachers examples—not exhaustive listings—of assessment items that reveal what students know and can do in mathematics, that pinpoint areas of strengths and weakness in students' mathematical knowledge, and that shape teachers' curricular and instructional decisions toward the goal of maximizing all students' understanding of mathematics.

Preface

The National Council of Teachers of Mathematics asked our task force to compile a rereresentative sample of assessment items that support *Principles and Standards for School Mathematics* (NCTM 2000). This book, one of four in the series, focuses on classroom assessment in grades 6–8. The other three books, for teachers in prekindergarten–grade 2, grades 3–5, and grades 9–12, also contain practical examples and samples of student work aligned with the NCTM Standards. Each of the books contains multiple-choice, short-response, and extended-response questions designed to help classroom teachers identify problems specifically related to certain of the NCTM Standards and Expectations. Matrices with this information are contained in the appendix.

NCTM's *Assessment Standards for School Mathematics* (1995) tells us that classroom assessment should—

- provide a rich variety of mathematical topics and problem situations;
- give students opportunities to investigate problems in many ways;
- question and listen to students;
- look for evidence of learning from many sources;
- expect students to use concepts and procedures effectively in solving problems.

Our collection of examples was compiled from many sources including state and provincial assessments. We know that standardized assessment has a major impact on what educators do in the classroom. Because most formal assessments include multiple-choice items, we have included them in this Sampler. Owing to the limited amount of information to be gleaned from most multiple-choice items, we have added an "explain your thinking," "justify your solution," or "how do you know?" component to most multiple-choice items. We believe that if students are going to be prepared to answer multiple-choice questions on formal assessments, they need classroom experience in answering this type of item, but we also want to be sure that students can support their answers by showing their work.

We have included a variety of rubrics as examples of how extended-response questions might be scored. We believe that students who know in advance how their answers will be evaluated will strive to meet the expected criteria; we realize, however, that for many assessment instruments, students are not privileged to this information. For classroom assessment, though, we believe that students should be given the rubric as a component of the assessment.

We encourage you to use these items with your students and hope that you find the bibliography and resources sections useful as you work toward extending your own class-room repertoire of assessment items.

Acknowledgments

The editors and writing team wish to thank the educators listed below for their suggestions, contributions of student work, reviews, and general assistance.

Katie Aspinwall

Larisa Ballard

Ann Bartosh

Margaret Bondorew

Cathy Brown

Jack Carrier

Patricia Cascioli

Marjorie Claytor

Beth Cole

Bill Collins

Janet Dandurand

Janie Dubé

Carol Findell

Ann Frederickson

David Goverman

Shannon Hager

Jane Hanna-Smith

Barbara Jackson

Sherry Kamm

Judith Keeley

Mark Lambke

Barbara Littlejohn

Allison Lokey

Sara Mischinik

Elizabeth Niemala

Brad Parker

Mattye Pollard-Cole

Beth Remme

Lyle Rogers

Edith Roos

Ellen Rose

Deborah Rubinstein

Richard Scheaffer

Sue Stewart

Eric Weiss

Nancy Wright

Lee Zimmer

About This Series

An emphasis on assessment, testing, and gathering evidence of student achievement has become an educational phenomenon in recent years. In fact, we can fairly say that assessment is driving many educational decisions, including grade placement, graduation, and teacher evaluation. With that influence in mind, educators need to use good assessment material as an essential tool in the teaching and learning processes. Good problems are those that are mathematically rich, can be solved in multiple ways, promote critical thinking, and can be evaluated in a consistent manner—that is, teacher X and teacher Y would be likely to evaluate a problem in the same manner with the appropriate rubric.

Assessment is actually only one of three major considerations in the processes of teaching and learning. As such, assessment must be viewed in conjunction with curriculum and instruction. Just as a curriculum aligned with standards can guide instructional decisions, so too can assessment guide both instructional and curricular decisions. Therefore, items designed to assess specific standards and expectations should be incorporated into the classroom repertoire of assessment tasks.

In its *Assessment Standards for School Mathematics* (*Assessment Standards*), the National Council of Teachers of Mathematics (NCTM 1995) articulated four purposes for assessments and their results: (1) monitoring students' progress toward learning goals, (2) making instructional decisions, (3) evaluating students' achievement, and (4) evaluating programs. Further, the Assessment Principle in *Principles and Standards for School Mathematics* (*Principles and Standards*) states that "assessment should not merely be done to students; rather it should be done for students" (NCTM 2000, p. 22). We have included a variety of rubrics in this series to assist the classroom teacher in providing feedback to students. Often, if students understand what is expected of them on individual extended-response problems, they tend to answer the questions more fully or provide greater detail than when they have no idea about the grading rubric being used.

This series was designed to present samples of student assessment items aligned with *Principles and Standards* (NCTM 2000). The items reflect the mathematics that all students should know and be able to do in grades prekindergarten–2, 3–5, 6–8, and 9–12. The items focus both on students' conceptual knowledge and on their procedural skills. The problems were designed as formative assessments, that is, assessments that help teachers learn how their students think about mathematical concepts, how students' understanding is communicated, and how such evidence can be used to guide instructional decisions.

The sample items contained in this publication are not a comprehensive set of examples but, rather, just a sampling. The problems are suitable for use as benchmark assessments or as evaluations of how well students have met particular NCTM Standards and Expectations. Some student work is included with comments so that teachers can objectively examine a particular problem; study the way a student responded; and draw conclusions that, we hope, will translate into classroom practice.

This series also contains a chapter for professional development. This chapter was developed with preservice, in-service, and professional development staff in mind. It addresses the idea that by examining students' thinking, teachers can gain insight into what instruction is necessary to move students forward in developing mathematical proficiency. In other words, assessment can drive instructional decisions.

NCTM's *Assessment Standards* (1995) indicates that (a) assessment should enhance mathematics learning, (b) assessment should promote equity, (c) assessment should be an open process, (d) assessment should promote valid inferences about mathematics learning, and (e) assessment should be a coherent process. This series presents problems and tasks that, when used as one component of the assessment process, help meet those Assessment Standards.

Introduction: About This Book

THE PURPOSE of this book is to provide the classroom teacher with examples of assessment items for grades 6–8 that are specifically and purposely aligned with the recommendations in *Principles and Standards for School Mathematics* (NCTM 2000). The varied sample assessment items collected here are referenced with the Content Standards and Expectations set forth in that document. The Process Standards–Problem Solving, Reasoning and Proof, Communication, Connections, and Representation– addressed by the assessment items are indicated in the items matrices in the appendix. Classifying an item as targeting a particular Expectation was not always an easy task, because good items address multiple, interrelated Expectations. Therefore, we have listed some items with multiple Standards and Expectations.

Students are being assessed on the state level with a preponderance of multiple-choice items, so we have chosen samples of multiple-choice items that we think have the potential to help teachers make informed instructional decisions. The distractors that students choose in answering multiple-choice items can often foster insight into students' misconceptions. Moreover, correct and incorrect responses to multiple-choice items can supply teachers with information that is useful in planning instruction. We have therefore included explanations for the distractors in many items.

We have added "teacher notes" to suggest ideas for making multiple-choice items more meaningful, either (a) by asking students to justify their answers or explain why they made their choice or (b) by removing the distractors, requiring students to produce the solution and justify their answers. We have also made suggestions for altering, extending, or modifying some of the items.

We have also chosen short-answer and extended-response items designed to give students opportunities to demonstrate their skills and understanding. An important outcome of assessment is the determination of how well students communicate mathematically and whether they use multiple representations. These skills are better assessed in short-answer or extended-response formats.

The professional development chapter is a guide for in-service and preservice teachers. We present suggestions for understanding and using levels of complexity for particular problems, using and adapting multiple-choice items, using assessment tasks as an in-service topic, using scoring rubrics, using technology in assessment, and designing universal assessment items. The professional development chapter is designed to be used by both in-service and preservice teachers as well as by professional development providers.

We hope that you find this compilation of problems and items interesting and helpful as you align your curriculum and assessment with the Standards and Expectations. The student work included in this Sampler is a reminder that students do not always think and reason about mathematical ideas as we teachers think they do. Rather, the student work illustrates the need for us to rethink how we can engage our students in developing a deeper understanding of mathematical ideas. Using actual student work can serve as a powerful vehicle for helping us realize the need to interview students to determine what they are thinking, guage their level of understanding about a concept, and guide our instructional decisions.

1

Number and Operations

N THE middle grades, students are expected to develop a deeper understanding of fractions, decimals, percents, and integers and to demonstrate proficiency in solving problems involving them. *Principles and Standards for School Mathematics* (NCTM 2000) identifies the importance of giving middle schoolers extensive experience with ratios, rates, and percents as a foundation for developing their understanding of proportionality. In addition, students in the middle grades should continue to refine their understanding of the four basic operations and how they operate with fractions, decimals, percents, and integers.

Middle school students are expected to deepen their understanding of rational numbers. They should be able to demonstrate fluency in working with different representation for fractions, decimals, and percents and to use them meaningfully. They should have multiple and varied experiences solving problems with a variety of models, including but not limited to fraction strips, number lines, area models, and concrete objects.

Students in grades 6–8 should also continue their work with whole numbers in a variety of problem-solving settings. They should deepen their understanding of very large and very small numbers. Students at this level are expected to develop an understanding of scientific notation and to use it appropriately to solve problems.

All students in the middle grades should demonstrate the ability to compute efficiently and accurately with fractions, decimals, and integers. They should understand when an estimate is more appropriate than the exact answer and how to choose the computational method that best fits a given problem situation.

The problems included in this chapter range in complexity and difficulty. The goal of the set is to show the breadth of the number strand. The examples can in no way be considered inclusive of all topics.

Number and Operations Assessment Items

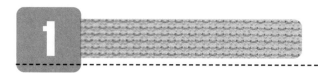

Mrs. Torres announced that 80% of the girls were interested in taking the class trip, whereas only 70% of the boys were interested. Carlos exclaimed, "That means that 75% of the class is interested in going on the trip." Is Carlos correct? Explain.

About the mathematics: This item involves understanding percent as a proportion and combining percents with possibly different bases.

Solution: Carlos is incorrect. One strategy is to make up a number of girls in the class and a number of boys in the class, for example, 10 girls and 18 boys. Then calculate 80% of 10 girls, giving 8 girls, and 70% of 18 boys, giving 12.6 boys. The total of students going on the field trip in this example is 20.6 students. But 75% of the class of 28 students would be 21 students, a total that is close to, but not equal to, 20.6 students. If larger numbers are used, the difference in the percentages of students is more obvious. For instance, 100 girls and 400 boys would result in 80% of 100 girls, or 80 girls, plus 70% of 400 boys, or 280 boys, giving a total of 360 students. But 75% of the 500 total students is 375 students.

About the student work: Students need to realize that the percentage of the whole group depends on the sizes of the groups. They also need to realize that they must deal in whole numbers; accordingly, they should demonstrate how they interpret remainders

Student Work

Student Response A

This student's response indicates a complete understanding of the problem.

No, it depends on the amount of boys and girls.

Student Response B

Student B may understand that the percentage of the whole group depends on the sizes of the groups, or may simply believe that the actual numbers are necessary to solve the problem.

Not neccessarily, the amount of boy and girls in the class is not given so given only this information there would be no way of knowing.

Student Response C

The work of student C is typical of students' incorrect responses. The student does not show an understanding of percents.

Yes Because 75% is the Averages of 80% & 70%

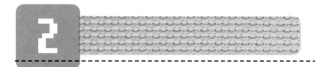

On Bill's football squad, $^1/_3$ of the players walk to practice and 25% are driven by their parents. The remaining 15 players take the bus. How many members are on the football squad?

A. 48
B. 24
C. 36
D. 30

About the mathematics: Students are asked to find the whole when given the part; they need to realize that the parts may be given in both fractions and percents.

Solution: We know that $^1/_3$ of the students walk and that $^1/_4$ are driven, thus accounting for $^1/_3 + ^1/_4$, or $^7/_{12}$, of the team. Therefore, $^5/_{12}$ of the students take the bus. If $^5/_{12}$ of the students is 15 students, then $^1/_{12}$ of the students is 3 students, meaning that 12×3, or 36, students are on the team.

Distractor rationales

   (a) A number greater than $15 + 25$
   (b) A number with 3 and 4 as factors
   (c) Correct
   (d) $^2/_3$ of 30 is 10, $^1/_4$ of 20 is 5, $30 - 10 - 5 = 15$.

Teacher note: The problem could be presented with all fractions or all percents. Asking students to make a visual representation of the team may help some solve the problem.

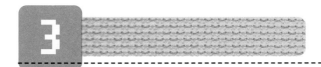

3

Friday was pajama day at Washington Elementary School. Ms. Wong's class has 20 students, and 14 of them were wearing pajamas. Mr. Locke's class has 28 students. On that day, Mr. Locke's class had 21 students wearing pajamas.

a. Which class had the higher percent of students wearing pajamas? Show your work.
b. Ms. Khan's class has 25 students. At least how many students would have to be wearing pajamas for her class to win the spirit award given to the class with the most people participating?

> **Source:** Adapted from Balanced Assessment for *More or Less*, Mathematics in Context (New York: Holt, Rinehart & Winston, © 1998 by Encyclopædia Britannica)
>
> **About the mathematics:** Students find percents given a part and the whole, compare percents, find a part given the percent and the whole, and reason about a situation involving percents and not whole numbers.
>
> **Solution**
>
> A: Mr. Lock's class; solution strategies will vary. One possible method is using a ratio table, as follows:
>
> Ms. Wong's class has 70%:
>
> | p.j. wearing | 14 | 7 | 70 |
> |---|---|---|---|
> | total | 20 | 10 | 100 |
>
> Mr. Locke's class has 75%:
>
> | p.j. wearing | 21 | 3 | 75 |
> |---|---|---|---|
> | total | 28 | 4 | 100 |
>
> B. Ms. Khan's class would have to have at least 19 people in pajamas. The percent of her class in pajamas must be more than 75%. Because 75% of 25 students is 18.75 students, at least 19 of them would have to be wearing pajamas.

Teacher note: The particular values were chosen to make the conversion to percents possible without a calculator.

About the student work: Students' strategies for finding the percents for each class may reveal their understanding of the concept of percent. Likewise, strategies for part b may reveal how flexibly students understand percents.

Student Work

Student Response

In this example, the student clearly understands the connection between decimals and percents and has used the fraction equivalent of the percent. The work is checked at the end.

$14/20 = $ ~~.~~ 0.7 or 70% for Ms. Smith's class

$21/28 = 0.75$ or 75% for Mr. Locke's class

a. Mr. Locke's class

b. 25% × 4 = 6.25 for 25%
   6.25 × 3 = 18.75 for 75% ← round up to get (19)
   $19/25 = $ ~~.~~ .76 or (76%)

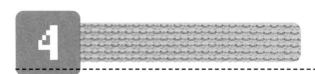

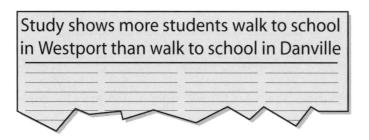

Study shows more students walk to school in Westport than walk to school in Danville

Here are the data provided in the article pictured on page 8.

	Percent of Students Who Walk to School
Westport	70%
Danville	50%

Explain how it *could* be possible that a greater number of students walk to school in Danville even though a greater percent of students walk to school in Westport.

> **About the mathematics:** Students need to understand that percent alone cannot be used to compare two different populations.
>
> **Solution:** If Danville is larger than Westport, then more students could possibly walk in Danville. For example, if Westport has 100 students and Danville has 200 students, then 70 students would walk in Westport and 100 would walk in Danville.
>
> **Teacher note:** This problem addresses the fact that a percent provides a relative comparison rather than an absolute one.

Standard: Understand numbers, ways of representing numbers, relationships among numbers, and number systems

Expectations: Work flexibly with fractions, decimals, and percents to solve problems; compare and order fractions, decimals, and percents efficiently and find their approximate locations on a number line

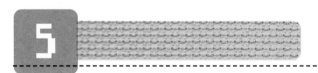

Name two fractions that come between

$$3/_5 \text{ and } 4/_5.$$

Justify your answer in two different ways.

About the mathematics: This item involves reasoning about the density of rational numbers.

Solution: Answers will vary. Students may use equivalent fractions, a number line, or an illustration to support their answers. Some possible solutions include $7/10$, $13/20$, $14/20$, and $15/20$.

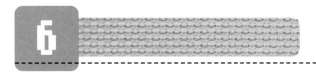

Which of the following fractions is larger,

$$8/9 \text{ or } 11/12?$$

Explain how you can determine which fraction is larger *without* changing the fractions to decimals or finding a common denominator.

About the mathematics: This item involves reasoning about fractions and explaining strategies for comparing fractions.

Solution: The larger fraction is $11/12$. In the instance of $8/9$, only nine parts compose the whole. In $11/12$, twelve parts compose the whole and the twelve parts must be smaller than the nine parts. If you divide the same whole into different-sized parts, then the greater the number of parts, the smaller each part must be. A $1/12$ piece is smaller than a $1/9$ piece, so a smaller piece of the whole is missing when $11/12$ remain than when $8/9$ remain, making $11/12$ the larger of the two fractions.

About the student work: Most students realized that $1/9$ was larger than $1/12$ but had a difficult time writing a sentence using that knowledge to logically explain why $11/12$ therefore had to be the larger fraction.

Student Work

Student Response A

The work demonstrates that the student has a good understanding of the relative size of the unit fractions and can apply it to the problem.

$\frac{11}{12}$ is bigger because $\frac{1}{12}$ is less than $\frac{1}{9}$.

Student Response B

This student has a good understanding of the relative size of the unit fractions but has a difficult time expressing it in a sentence.

$\frac{11}{12}$ is larger because it is missing

a smaller piece if you tryed to make it

a whole than $\frac{8}{9}$

Student Response C

The answer is incorrect. The student does seem to know the relative size of the unit fractions or at least that $\frac{1}{9}$ is bigger than $\frac{1}{12}$. More questioning is necessary to gain further understanding of the student's knowledge.

with $\frac{8}{9}$ the pieces get bigger
because you don't have to cut
them smaller to make more like $\frac{11}{12}$.

> **Standard:** Understand numbers, ways of representing numbers, relationships among numbers, and number systems

> **Expectations:** Develop meaning for percents greater than 100 and less than 1; understand and use ratios and proportions to represent quantitative relationships

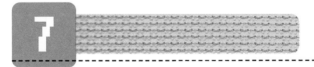

The following are pictures of a house that have been enlarged. The dimensions of the house in picture B are 150% of the dimensions of house A. The dimensions of the house in picture C are 150% of the dimensions in picture B. Some of the dimensions are provided. Determine the lengths of the missing dimensions *x, y,* and *z*

Note: Drawings are not to scale.

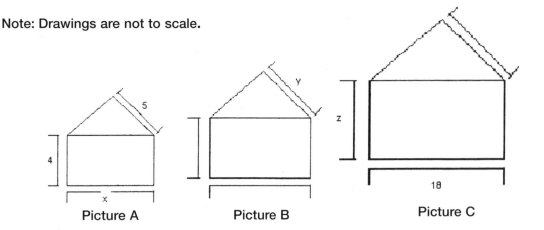

Picture A   Picture B   Picture C

Which of the following shows the length of *x, y,* and *z* in order from the shortest length to the longest length?

A) *x, y, z*
B) *y, x, z*
C) *x, z, y*
D) *y, z, x*

About the mathematics: Students need to understand percent increase in a geometric context, determine input that provides a given result, and order variables on the basis of their values

Solution: On picture C the base is 18, so the base on picture B must be 18 ÷ 1.5, or 12. Then *x* would be 12 ÷ 1.5, or 8. To get *y*, scale up from picture A: 5 × 1.5 = 7.5. Finally, *z* is a two-step enlargement: 4 × .5 = 6 for picture B, and *z* is 6 × 1.5, or 9. Thus, *x* = 8, *y* = 7.5, and *z* = 9. The correct order is *y, x, z*.

Distractors
   A) Ordered by same dimension
   B) Correct
   C) Length, height, diagonal
   D) Visual impression from drawing

Teacher note: More could be discovered about the student's ability to work with this concept by changing the question to a short-answer question and requiring all work to be shown. The problem was designed with the specific requirement to apply the 150 percent increase in both directions. The decision was made to not draw the houses to scale, so that students could not solve the problem using measurement. The students could be asked to draw the houses to scale, but then the problem overlaps both number and measurement. This problem can also be thought of as dealing with similarity, thus overlapping with geometry.

Student Work

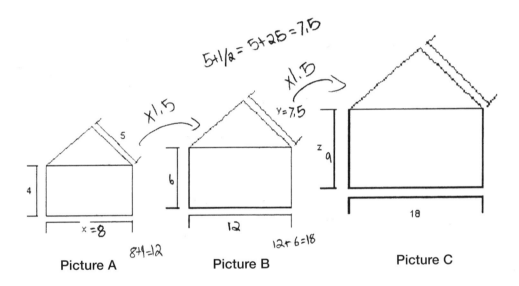

Picture A   Picture B   Picture C

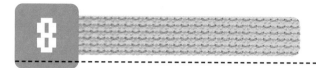

Plato's Pizza introduced a new pizza called the Super Giant to compete with Euclid's 2-pizzas-for-the-price-of-1 offer. The Super Giant is two 1-square-foot pizzas put together. Plato's ad claims that the Super Giant is 25% larger than two of Euclid's 12-inch round pizzas. A Super Giant costs $10.99. Two 12-inch round pizzas from Euclid's cost $10.88.

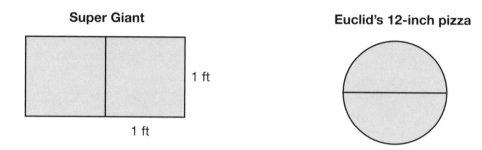

Super Giant

1 ft

1 ft

Euclid's 12-inch pizza

a   Which offer gives you more pizza for your money?
b.   Is the Super Giant 25% larger than two 12-inch round pizzas from Euclid's? If so, prove it. If not, use percents to show how they really compare.

> **Source:** Adapted from *Connected Mathematics Question Bank* (Upper Saddle River, N.J.: Prentice Hall, 2002)
>
> **About the mathematics:** Students use porportion to make a quantitative comparison between two offers, both of which refer to a 12-inch measurement.
>
> **Solution**
>
> a.   The Super Giant pizza has an area of $2 \times 12 \times 12$, or approximately 288, square inches. Calculate 288 in^2/$10.99, giving 26.2 square inches per dollar. By comparison, two Euclid's pizzas have an area of $2 \times \pi \times 36$, or 226, square inches. Calculate 226 / $10.88, giving 20.8 square inches per dollar. Thus the Super Giant pizza gives you more pizza for your money.
>
> b.   Since 288/226 = 1.27, the Super Giant is actually about 27 percent larger than two Euclid's pizzas.

Standard: Understand numbers, ways of representing numbers, relationships among numbers, and number systems

Expectations: Understand and use ratios and proportions to represent quantitative relationships; develop an understanding of large numbers and recognize and appropriately use exponential, scientific, and calculator notation

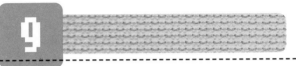

In astronomy, a light year is 5,880,000,000,000 miles.

a   Write the number of miles in 1 light year in scientific notation.

b. Write the number of miles in 100 light years in scientific notation. Explain your reasoning.

c. The distance between two stars is 3×10^{11} miles. Without using your calculator, estimate how many light years apart the stars are. Explain how you found your answer.

Source: *Connected Mathematics Question Bank* (Upper Saddle River, N.J.: Prentice Hall, 2002)

About the mathematics: This item involves working with scientific notation and reasoning about scientific notation.

Solution

    a. 5.88×10^{12} miles

    b. 5.88×10^{14} miles. If 1 light year is 5, 880,000,000,000 miles, then 100 light years is 588,000,000,000,000, meaning that the decimal point must be moved two additional places to the right.

    c. One light year can be rounded to 6×10^{12} miles. Dividing 6×10^{12} miles by 3×10^{11} miles results in 2×10^{1}, or $2 \times 10 = 20$, light years apart.

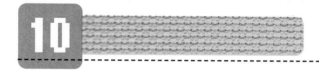

Radio signals travel at the speed of light. The speed of light is 1.86×10^5 miles per second. Without using a calculator, how could you calculate the number of seconds a radio signal would take to travel from the sun to earth if the earth is 9.14×10^7 miles from the sun? About how long will the signal take to travel to earth?

> **About the mathematics:** The solution involves applying scientific notation and using the rule of exponents under division.
>
> **Solution:** Because the unknown is time given the distance and the rate, the time can be found by dividing the distance by the rate: $(9.14 \times 10^7 \text{ mi})/ (1.86 \times 10^5 \text{ mi/sec}) = 4.91 \times 10^2$, or 491, seconds.

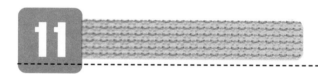

Jeremy was boasting to Tabitha that he could determine the ones digit in any problem in which 2 was raised to any power. He claimed he could do so without a calculator. Tabitha was sure this task was impossible, so she quizzed Jeremy. Here are the results:

$$2^{100} \quad \ldots \ldots \ldots \ldots \ldots \quad \text{last digit} = 6$$
$$2^{93} \quad \ldots \ldots \ldots \ldots \ldots \quad \text{last digit} = 2$$
$$2^{91} \quad \ldots \ldots \ldots \ldots \ldots \quad \text{last digit} = 8$$

- Explain how Jeremy is able to determine the ones digit so easily.
- Could you use a system like Jeremy's for powers of 3 or other numbers? If so, explain how. If not, explain why not.

> **Source:** Adapted from Vermont Department of Education. http://www.tea.state.tx.us/student.assessment/resources/release/taks/2003/ gr10taksmath.pdf, item 23 (accessed March 22, 2005).
>
> **About the mathematics:** The solution involves working with powers of 2.

Solution: By calculating the first five results of raising 2 to sequential powers, one can see a pattern: $2^1 = 2$, $2^2 = 4$, $2^3 = 8$, $2^4 = 16$, $2^5 = 32$. In this pattern appears a cycle in which the units digit repeats for every fourth value. Jeremy may have noticed that if he *divides the exponent* by 4, he will get remainders of 0, 1, 2, or 3. If no remainder results, then the units digit is a 6. If the remainder is 1, the units digit is a 2; if the remainder is 2, the units digit is 4; and if the remainder is 3, the units digit is 8. When working with powers of 3, the pattern is

$$3^1 = 3, 3^2 = 9, 3^3 = 27, 3^4 = 81, 3^5 = 243.$$

In this instance the pattern also repeats for every fourth value. But when the exponent is *divided by 4* and no remainder results, the units digit is 1; a remainder of 1 indicates a units digit of 3; a remainder of 2, a units digit of 9; and a remainder of 3, a units digit of 7. This pattern will work for other bases raised to powers.

Standard: Understand numbers, ways of representing numbers, relationships among numbers, and number systems

Expectation: Use factors, multiples, prime factorization, and relatively prime numbers to solve problems

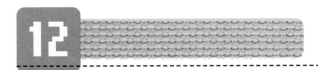

Two radio stations are playing the number-one hit song "2 B Nice 2 U." WBCU plays the song every 24 minutes. WFSC plays the song every 9 minutes. Both stations play the song at 1:30 p.m.

a. When is the next time the stations will play the song at the same time?
b. When is the next time they will both play the song on the half hour?

Source: Adapted from *Connected Mathematics Project,* "Prime Time" (Upper Saddle River, N.J.: Prentice Hall, 1996)

About the mathematics: This item involves the application of greatest common multiples.

Solution

   a. In 72 minutes, or at 2:42 p.m.
   b. At 7:30 p.m.

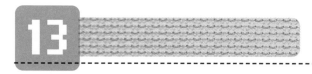

Meredith is planning a Halloween party. She has 36 prizes and 24 balloons. What is the most number of children she can invite to her party so that each child gets an equal number of prizes and an equal number of balloons. She does not want any prizes or balloons left over. Explain your answer.

Source: Adapted from *Connected Mathematics Project*, "Prime Time" ((Upper Saddle River, N.J.: Prentice Hall, 1996)

About the mathematics: Students are required to apply greatest common factors to a contextual problem.

Solution: The greatest common factor of 24 and 36 is 12. If Meredith has 12 children at her party, each child would receive 3 prizes and 2 balloons.

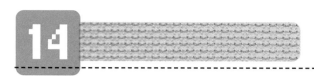

a. Write the largest string of factors for the product

720.

b. Write your answer using exponents.

About the mathematics: This item involves prime factorization.

Solution

   a. $720 = 2 \times 2 \times 2 \times 2 \times 3 \times 3 \times 5$
   b. $2^4 \times 3^2 \times 5$

Standard: Understand numbers, ways of representing numbers, relationships among numbers, and number systems

Expectation: Develop meaning for integers and represent and compare quantities with them

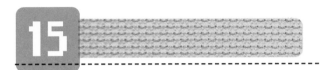

When flying from Minneapolis to Seattle, here is the flight information.

Flight	Date	Itinerary	Time
NW1607	7/26	Depart Minneapolis Arrive Seattle	11:30 A.M. 1:00 P.M.

When flying from Seattle to Minneapolis, here is the flight information.

Flight	Date	Itinerary	Time
NW 075	7/31	Depart Seattle Arrive Minneapolis	11:45 A.M. 5:15 P.M.

Explain why the flight from Seattle to Minneapolis appears to take so much longer than the flight from Minneapolis to Seattle.

Source: Adapted from *Operations*, Mathematics in Context (New York: Holt Rinehart & Winston, © 1998 by Encyclopædia Britannica, problem 3, p. 12)

About the mathematics: The solution involves reasoning with positive integers using time zones.

Solution: Between the Central Time Zone and the Pacific Time Zone is a two-hour time difference.

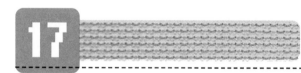

Suppose you are in a building in which the floors are numbered from 0 to 15. The building has an underground parking garage with 10 levels, which are numbered from –1 to –10. Which floor is farther from floor–2?

a. Floor 7 or floor –10
b. Floor 2 or floor –8
c. Floor 1 or floor –5

> **Source:** Adapted from *Connected Mathematics Project*, "Accentuate the Negative" (Upper Saddle River, N.J.: Prentice Hall, 1996)
> **About the mathematics:** This item involves reasoning with integers.
> **Solution**
> a. Floor 7 is farther from floor – 2.
> b. Floor –8 is farther from floor –2.
> c. They are both 3 floors from floor –2.

Standard: Understand meanings of operations and how they relate to one another

Expectation: Understand the meaning and effects of arithmetic operations with fractions, decimals, and integers

Sailing against the current caused a boat to have a net loss of –11.7 miles in its forward progress over 4.5 hours. What was its net loss or gain per hour?

a. –2.6 miles
b. –7.2 miles
c. –16.2 miles
d. –52.64 miles

Source: Nova Scotia Department of Education, provincial assessment released item

About the mathematics: The solution involves division of integers.

Solution: The correct response is a.

Distractors

    b. Subtracting 4.5

    c. Adding −4.5

    d. Multiplying by 4.5

Standard: Understand meanings of operations and how they relate to one another

Expectation: Understand the meaning and effects of arithmetic operations with fractions, decimals, and integers

---

Standard: Understand numbers, ways of representing numbers, relationships among numbers, and number systems

Expectation: Work flexibly with fractions, decimals, and percents to solve problems

---

18

If $x\%$ of 425 is 598, then what must be true?

a. $25 \leq x \leq 65$.

b. $66 \leq x \leq 110$.

c. $110 \leq x \leq 150$.

d. $151 \leq x \leq 200$.

About the mathematics: Students need to estimate the percent taken given the original and the result.

Solution: The correct response is c. This problem can be thought of as representing about 600/425, which is equivalent to about 100/70, or approximately 140 percent. Only response c includes the numbers that occur near 140.

Teacher note: The format of the problem encourages estimation.

About the student work: Students need first to realize that in the instance of an increase, x must be greater than 100.

Student Work

Because 598 is bigger than 425 it must be more than 100°. 50% of 425 is over 200, so 150% is over 600 so the answer is C.

19

Graham received a coupon in the mail for a 30% discount on any pair of jeans at Cool Jeans Hut. When he arrived at the store, the store was offering an additional 20% discount off all purchases. Graham picked out a pair of Jeans for $40.00. The clerk added the 30% and 20% discounts and charged Graham $20. Later, the manager of the store told the clerk that he had undercharged Graham.

a. By how much did the clerk undercharge Graham?
b. Explain to the clerk what he did wrong, and what he should do in the future when more than one discount is applied on a purchase.

About the mathematics: Students must understand how to calculate sequential percents.

Solution

a. $1.20

b. The clerk added the percents when in fact the second discount should have been applied to a smaller total. The total discount should have been 44%. In the future the clerk could first find one discount, then calculate the second discount. Alternatively, the clerk could multiply the two discounts, then multiply times the original amount.

Teacher note: Some students may not know that retail discounts work in this way; however, the problem clearly states that a mistake has been made. Part b often reveals students' misconceptions about percents

Student Work

Student Response A

Student A's work shows a complete understanding of the problem.

Part a

He will take out $1.20 because 20% was taken off so it would be $16. Then 30% would be taken off so it would be $11.20. 11.20 - 10.00 = 1.20

Part b

The problem was that he just added the 20% and 30% together to take off 50% when he should have done one discount, then the other. 44% shall have been taken off

Student Response B

The work reveals that student B is not clear about what constitutes the whole and thus has made errors. Calculation errors have led the student to believe that the order of the discounts matters.

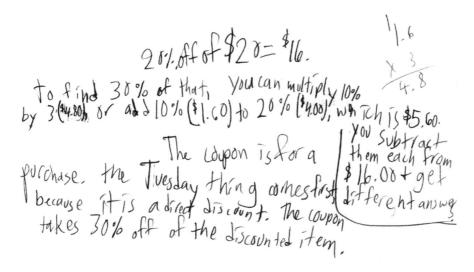

Student Response C

This work is also incorrect, indicating that student C does not understand that the order of the discounts does not matter.

Part a

$1.20 because it should have costed 11.20.

20
.3
———
6.0
14
.2
———
2.4

Part b

the number is diferent after you alreacd take some % of so you can't just add the % toget and then take off the discount.

11.20   30% then 20% or
20% then 30%

Abigail Adams Middle School students are trying to make the largest submarine sandwich on record. They made an Italian cold-cuts sandwich that was $12\frac{3}{4}$ feet long. After realizing that they had not set a record, they decided to divide the sandwich into smaller portions to share with other students.

a. If each portion was $\frac{1}{2}$ foot long, how many students would get a portion?
b. If each portion was $\frac{3}{4}$ foot long, how many students would get a portion?

Justify your answer. Be sure to show all your work.

> **About the mathematics:** This item involves the division of fractions and mixed numbers in a contextual situation.
>
> **Solution**
> a. $12\frac{3}{4} \div \frac{1}{2}$ could be simplified by thinking that $12\frac{3}{4}$ students would get a share if the portions were 1 foot long, so twice as many students will actually get a share because each share is only $\frac{1}{2}$ foot long. Calculate $12(2) = 24$ and $\frac{3}{4}(2) = 1\frac{1}{2}$; $25 + 1\frac{1}{2} = 25\frac{1}{2}$. So 25 students will get a sandwich, with $\frac{1}{2}$ of a $\frac{1}{2}$-foot portion left over.
> b. $12\frac{3}{4} \div \frac{3}{4}$ could be thought of in this way: If each student was getting $\frac{1}{4}$ of a 1-foot portion of the sandwich, 48 students would share the 12-foot length and 3 students would share the $\frac{3}{4}$-foot length. But because each student gets $\frac{3}{4}$ of a 1-foot portion, not $\frac{1}{4}$ of a 1-foot portion, calculate $48 \div 3 = 16$ and $\frac{3}{4} \div 3 = 1$; therefore, a total of $16 + 1$, or 17, students will share the sandwich.
>
> **Teacher note:** Regional terms for a submarine sandwich include hoagie, hero, grinder, and poor boy.

Standard: Understand meanings of operations and how they relate to one another

Expectation: Use the associative and commutative properties of addition and multiplication and the distributive property of multiplication over addition to simplify computations with integers, fractions, and decimals

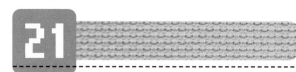

When shopping during a two-hour sale for $1/2$ off everything in the store sale, Jill selected three items originally priced $12.66, $95.98, and $47.34. Which of the following would *NOT* provide the correct total sale price?

a. $1/2 \times \$12.66 + 1/2 \times \$95.98 + 1/2 \times \$47.34$
b. $1/2(\$12.66 + \$95.98 + \$47.34)$
c. $1/2 \times \$12.66 + \$95.98 + \$47.34$
d. $1/2 \times \$60.00 + 1/2 \times \$95.98$

> **About the mathematics:** The solution involves using the distributive and associative properties of arithmetic.
> **Solution:** c

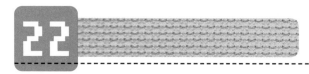

Create a set of data that includes 5 different temperatures. Your data must meet the following conditions:

*   The sum of the highest and lowest temperatures must be 0°.
*   The mean temperature must be –2°.

Explain how you know your data meet the two conditions.

Source: Nova Scotia Department of Education, provincial assessment released item
About the mathematics: This item involves using integers to represent situations.
Solution: Answers will vary. A sample might be –10, –9, 0, 5, 10.

Standard: Understand meanings of operations and how they relate to one another

Expectation: Understand and use the inverse relationships of addition and subtraction, multiplication and division, and squaring and finding square roots to simplify computations and solve problems

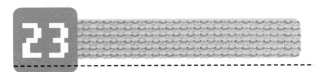

23

Write and solve a word problem that involves decimal numbers and can be solved in two ways: by multiplication or by addition. Show BOTH ways of solving your problem.

Source: Nova Scotia Department of Education, provincial assessment released item
About the mathematics: Students must understand multiplication as repeated addition.
Solution: Answers will vary. Here is one example: What is the total cost of five CD-RW disks if they can be purchased individually for $0.99? To solve, either multiply (5 × $0.99 = $4.95) or add ($0.99 + $0.99 + $0.99 + $0.99 + $0.99 = $4.95).

Standard: Understand meanings of operations and how they relate to one another

Expectations: Use the associative and commutative properties of addition and multiplication and the distributive property of multiplication over addition to simplify computations with integers, fractions, and decimals; understand and use the inverse relationships of addition and subtraction, multiplication and division, and squaring and finding square roots to simplify computations and solve problems

Standard: Compute fluently and make reasonable estimates

Expectation: Develop and use strategies to estimate the results of rational-number computations and judge the reasonableness of the results

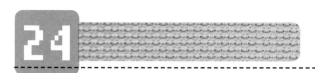

A calculator is suggested for the following problem.

When she was born, Krista Ann was given a gift of $1000. Her parents invested it in a college savings account that earns 3.2% annual interest.

a. How much money will Krista Ann have after 8 years? How did you determine this value?
b. How much money will she have after 16 years?
c. How long will it take to have $2000 in the account? How did you determine this value?

About the mathematics: The solution involves calculating compound interest.

Solution

a. $1286.58. Using a calculator, start with 1000 and then multiply by 1.032 eight times.

b. $1655.29

c. Between 25 and 26 years. Using a calculator, continually multiply by 1.032 until attaining a number bigger than 2000.

Teacher note: This problem is calculator dependent. Students may be tempted to double the answer from part a to find part b.

About the student work: Successful students understand the notion of compound interest. The most common error in part a was simply to calculate the interest for one year and then multiply by 8, ignoring the compounding nature of the accumulation.

Student Work

Student Response A

This student attempted to calculate compound interest but made calculation errors.

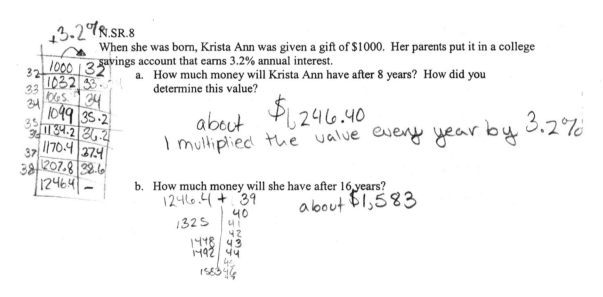

Student Response B

This student's work contains the most often seen error, lack of compounding.

a. How much money will Krista Ann have after 8 years? How did you determine this value?

~~1000.256~~

$1000 \times .032 \times 8 = 256$ ⚡

$\boxed{1256 \text{ \$}}$

b. How much money will she have after 16 years?

$0.032 \times 16 = 0.512 + 1000 = 1000.512$

$1000 \times .032 \times 16 =$

1576

c. How long will it take to have $2000 in the account? How did you determine this value?

close to 32 years

25

The map at the right shows some farms. The county charges a tax based on the size of each farm, not on the quality of the land, so all land is taxed at the same rate. The Joneses pay $5,250 per year in taxes. After the map was drawn, the Bakers purchased some property from the Petits and increased the size of their property to 125% of its size on the map. By approximately what percent will the Petit's tax bill decrease? Please make sure your solution strategy is clear.

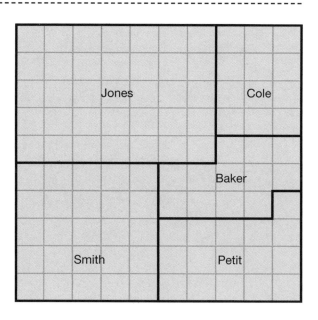

About the mathematics: Students estimate areas from a visual representation, change units to solve a problem, and calculate the percent of increase.

Solution: 18.75% (about 19% or 20%). This problem can be solved with or without the use of the tax price. Since the Joneses pay $5,250 and have 35 squares, we know that that the tax rate is $150 per square. The Bakers have 12 squares and increase their holdings to 125% of 12, or 15, squares, an increase of 3 squares. Therefore, the Petits reduce their land from 16 squares to 13 squares. The tax on 16 squares is 150 × 16, or $2400; the tax on 13 squares is $1950. The amount the Petits pay decreases $450. To find the percent of decrease, divide: 450/2400 = 18.75%. The Bakers have 12 squares. When they buy the land, they have 1.25 × 12, or 15, squares, meaning that they bought 3 squares. The Petits had 16 squares and now have 13. To find the percent of decrease, divide: $^3/_{16}$ = 18.75%.

Rubric (5 possible points)

1 point: Establishes the sizes of the farms

2 points: Calculates how much land changes hands

2 points: Shows correct calculation of 18.75 percent

Teacher note: This task is best suited for students working in groups. It is not designed as an on-demand task. Note that the answer remains the same regardless of the tax rate.

Student Work

Student Response A

The student giving this response did not use the tax rate.

Student Response B

This student figured the change in the tax and then figured the percent from that amount.

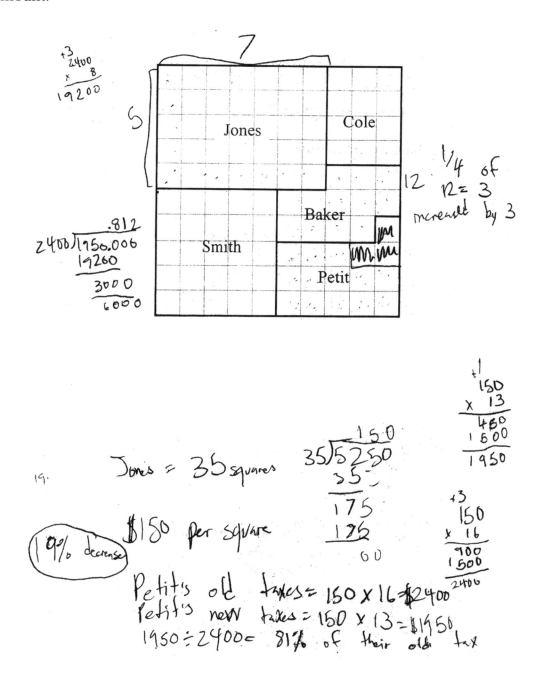

Standard: Compute fluently and make reasonable estimates

Expectation: Select appropriate methods and tools for computing with fractions and decimals from among mental computation, estimation, calculators or computers, and paper and pencil, depending on the situation, and apply the selected methods

26

The price of a gallon of gas has increased from $1.53 to $1.71. The newspaper headline reads "Price of gas up almost _____%." What number should go in the blank? Explain your answer.

About the mathematics: Given a price increase, students are asked to calculate the percent of increase.

Solution: A percent that falls in the 10%–20% range is acceptable because the problem states *almost*. To solve, calculate 1.71 ÷ 1.53 ≈ 1.12, meaning that the new price is 1.12 times the old price, or 112% greater; thus the increase is 12%. Or use mental mathematics to think as follows: The price difference is $0.18. A 10% increase would be 10% of $1.53, or just above $0.15. An additional 2% of $1.53 would be a little more than $0.03, so a 12% increase would be about $0.15 + $0.03 = $0.18. So about 12% is an acceptable response.

About the student work: Successful students are able to find the price change and then convert that amount to a percent of the initial price. The two examples that follow show slightly different calculations. A common error is to compare the change with the new price, getting an answer between 10 percent and 11 percent.

Student Work

Student Response A

This student started by using a division strategy but moved to using 10 percent and 1 percent.

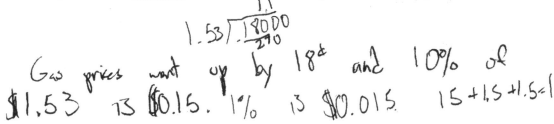

The price of a gallon of gas has gone from $1.53 to $1.71. The newspaper headline reads "Price of gas up almost ___12___ %" What number should go in the blank? Explain.

Gas prices went up by 18¢ and 10% of $1.53 is $0.15. 1% is $0.015. 15+1.5+1.5=1

Student Response B

This student appears to have estimated and then checked 12 percent.

The price of a gallon of gas has gone from $1.53 to $1.71. The newspaper headline reads "Price of gas up almost __12__ %" What number should go in the blank? Explain.

The number that should go in the blank is 12. You are adding 18¢ when you go from $1.53 to $1.71, so you know it's going to be just a little over 10%. Just take 12% which is 18.36¢, and you've got your answer.

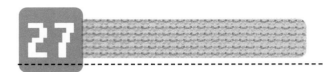

27

Two groups of tourists each have 60 people. If 75% of the first group and $2/3$ of the second group board buses to travel to a museum, how many more people in the first group board buses than in the second group? Explain your reasoning.

a. 2
b. 4
c. 5
d. 40
e. 45

> **About the mathematics:** Students are asked to compare different-sized parts of the same whole.
>
> **Solution:** Find 75% of 60, which is 45, and find $2/3$ of 60, which is 40; the difference is 5.
>
> **Teacher note:** This problem could be presented with only fractions or only percents.
>
> **About the student work:** Successful students calculated 75 percent of 60, usually by thinking of 75 percent as $3/4$, and found $2/3$ of 60; finally they subtracted the latter result from the former.

Student Work

Student Response A

Notice that although the notation is not completely clear in this work, one can readily see that the student calculated $3/4$ by finding $1/4$ and multiplying, then found $2/3$ in a similar way, and finally subtracted.

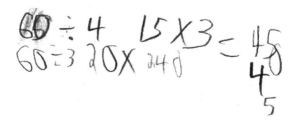

Student Response B

This student's use of a chart helps the organization of a disorganized solution. The student used a fraction bar to determine 75 percent, then calculated $^2/_3$, and finally subtracted.

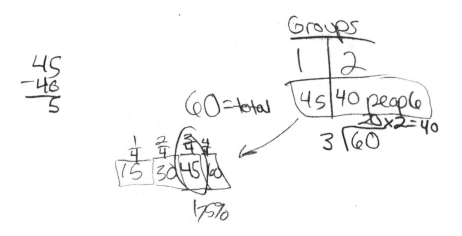

28

Tim needs to buy 4 new tires for his car. He found the following three ads in his local newspaper:

Steve's Tires	Tire Mart	Collins' Tire
Each tire: $63.55 Buy 3 and get one free!	Each tire: $46.25	Each tire: $56.75 20% off the total purchase

Where would Tim get the best deal? Show all your work, and explain how you decided which store would give him the best deal.

About the mathematics: This item involves comparing costs.

Solution

The total cost at Steve's tires will be 3 × $63.55, or $190.65.

The total cost at Tire Mart will be $46.52 × 4, or $186.08.

The total cost at Collins' Tire will be 0.8($56.75 × 4), or $181.60.

So Collins Tire has the cheapest price.

About the student work: Students figured each of the prices and then compared the results.

Student Work

Student Response

This student's work typifies the correct responses received.

Steves tires
$190.65

Tire mart
$185

Collins tire
$181.60

(Collins tire)

I took $63.55 × 3 to get steve's tires, Tire mart I took $46.25 × 4, Collins tire I took $56.75 × 4 and ÷ 5 to get $45.40 which is the discount then took $227 – 45.40 to get the answer

29

Dan's Discount Warehouse is having a half-price sale on every item in the store. Clearance items are marked down an additional 25%. Sydney plans to use a store coupon worth 25% off any item (even clearance items) in the store. Sydney says, "This means that I don't have to pay anything to buy a clearance item!" Do you agree with Sydney? Explain your thinking.

About the mathematics: The solution involves working with compound discounts.

Solution: In the correct answer, the student does not agree with Sydney. Their explanations may vary. Here are some sample explanations:

- Let us say that one item in the store has an original price of $20. Now it is on sale for half off that price, which is $10. If it was a clearance item, it would be marked down an additional 25% off. One calculation is as follows: 1/4 of $10 = $2.50; $10.00 – $2.50 = $7.50. When Sydney goes to use her 25% off coupon, she will save an additional $1.88 (i.e., 0.25 × $7.50 = $1.88). So the final sale price for that item would be $5.62.
- Whenever one finds a fractional part of a whole number, such as $\frac{1}{2}$ of 10, and then finds a fractional part of that new number, such as $\frac{1}{2}$ of 5, the answer will never come out to be 0: $\frac{1}{2} \times 10 = 5$, $\frac{1}{2} \times 5 = 2.5$, $\frac{1}{2} \times 2.5 = 1.25$, ...

About the student work: Students who agree with Sydney do not understand that in this problem involves a changing whole. Successful students realize that each successive discount is taken off the new (discounted) price.

Student Work

Student Response A

The student giving this response seems to have some understanding of the situation, but the explanation is not clear.

No I don't agree. Because it's takeing 25% off a lower number. So you'll get a lower discount.

Student Response B

This student provides a good solution, but without an example.

No, because first it is 50% off of the original price, then clearance items are 25% off the discount price, then the coupon would be 25% off that discount price.

Student Response C

This student provides a good solution and also includes an example.

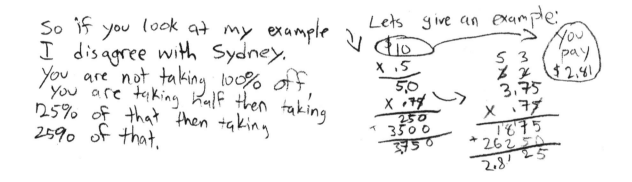

So if you look at my example I disagree with Sydney. You are not taking 100% off, you are taking half then taking 25% of that then taking 25% of that.

Lets give an example:

$10
× .5
5.0
× .75
250
3500
3750

5 3
3.75
× .75
1875
+26250
2.8125

you pay $2.81

Student Response D

The explanation in this response is typical of those given by students who do not see the changing whole.

I do agree with sydney because if everything is marked half price and the clearance is marked 25% off sydney is left with 25%. She has a coupon that is 25% so she won't have to pay any thing.

Standard: Compute fluently and make reasonable estimates

Expectation: Develop and analyze algorithms for computing with fractions, decimals, and integers and develop fluency in their use

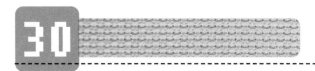

The city planetarium can seat 318 people. Anna is attending a show there that night, and hears that 225 people are in attendance. Anna wonders whether this number is correct. Which of the following methods could Anna use to see whether 225 seems reasonable?

Method 1: Look around to see whether about 2 out of every 3 seats are filled.
Method 2: Look around to see whether about two-thirds of the seats are empty.
Method 3: Look around to see whether the ratio of empty seats to filled seats is about 1 to 3.

   a. Method 1 only
   b. Method 2 only
   c. Methods 1 and 3 only
   d. Any of the methods could be used.

> **Source:** Nova Scotia Department of Education, provincial assessment released item
>
> **About the mathematics:** Students must understand estimation as being an approximation, not the rounding of the correct answer.
>
> **Solution:** The correct response is c.

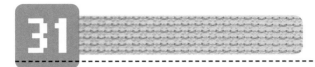

John is going to explain to Ben why $(-3) + (+5) = +2$. If ■ represents +1 and □ represents −1, draw a picture to show how John used these tiles to explain the answer to Ben.

Source: Nova Scotia Department of Education, provincial assessment released item

About the mathematics: Students use manipulatives to model computation with integers.

Solution

□ □ □		□ □ □ □ □		
	+	■ ■ ■ ■ ■	=	(crossed) ■ ■ ■ ■ ■
−3	+	0	=	+2

About the student work: The student work we reviewed suggests that most students have memorized a procedure for dealing with subtracting a negative integer; however, their work reveals little understanding of the concept or the use of the manipulative to assist in that understanding.

Student Work

Student Response A

The work of this student shows no understanding of the manipulative.

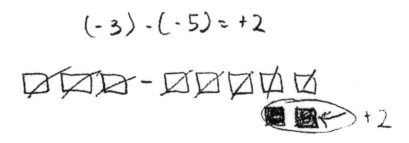

Student Response B

The work shows the algebraic solution but indicates no apparent understanding of why the negative value turns to a positive value.

2 because when you remove the brackets
The two negatives turn to positive so it becomes +5 but
-3 stays the same,

Student Response C

The work demonstrates a good understanding of the manipulative.

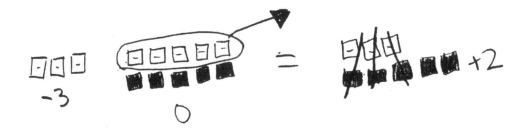

Student Response D

The student work evidences no understanding of the manipulative. The solution is algebraic, whereas the student has drawn a picture of the solution using the manipulative.

$$(-3) - (-5) = +2$$
$$(-3) + (+5) = +2$$

Standard: Compute fluently and make reasonable estimates

Expectation: Develop and use strategies to estimate the results of rational-number computations and judge the reasonableness of the results

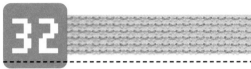

The product of $\left(-\frac{5}{8}\right) \times \left(-\frac{4}{3}\right)$ is closest to which number? Explain your reasoning.

a. $\frac{1}{2}$

b. 1

c. -1

d. 2

Source: Nova Scotia Department of Education, provincial assessment released item

About the mathematics: This item involves estimating the result of multiplication of rational numbers as well as the rules for multiplying integers.

Solution: The correct response is b. We know that $\frac{5}{8}$ is more than $\frac{1}{2}$ and that $\frac{4}{3}$ is more than 1 whole, so the estimate must be greater than $\frac{1}{2}$ and less than 1 whole. Since the problem involves $\frac{5}{8}$ of $1\frac{1}{3}$, the estimate should be closer to 1 than to $\frac{1}{2}$. Since a negative times a negative is positive, we can eliminate -1 as an answer.

Standard: Compute fluently and make reasonable estimates

Expectation: Develop, analyze, and explain methods for solving problems involving proportions, such as scaling and finding equivalent ratios

Joe made a scale drawing of a washing machine using the scale 1:3. His teacher, Mr. Eckhardt, wants to do a quick check to see whether Joe's drawing seems to be accurate. Mr. Eckhardt knows the ratio of the height to the width of the washing machine. Which of the following methods could he use to determine whether the dimensions of the washing machine on the drawing seem to be reasonable?

a. Check to see whether the ratio of the height to the width of the machine is the same on the drawing as it is on the real machine.
b. Check to see whether the ratio of the height to the width of the machine in the drawing is one-third the ratio on the real machine.
c. Check to see whether the ratio of the height to the width of the machine in the drawing is 3 times the ratio on the real machine.
d. Check to see whether the ratio of the height to the width of the machine in the drawing is 2 times the ratio on the real machine.

> **Source:** Adapted from Nova Scotia Department of Education, provincial assessment released item
> **About the mathematics:** This item requires a conceptual understanding of ratios.
> **Solution:** The correct response is a.

Algebra

STUDENTS in the middle grades should learn algebra both as a set of concepts and competencies tied to the representation of quantitative relationships and as a style of mathematical thinking for formalizing patterns, functions, and generalizations" (NCTM 2000, p. 223). Students at this level are expected to work regularly with algebraic symbols. As students develop their algebraic reasoning, they should be engaged in the use of tables, charts, graphs, and symbolic expressions containing variables. They should have multiple opportunities to recognize and generate equivalent expressions, solve linear equations, and use simple formulas.

In the middle grades, the study of patterns and relationships should focus on patterns that relate to linear functions. As students solve problems, they should be able to identify situations in which a constant rate of change is involved by examining tables, charts, graphs, and equations. They should have ample experiences in comparing data from two data sets that show linearity. They are expected to determine what the intersection of the two data sets represents, how to interpret the information that comes before or after the intersection of the data sets, and how a nonlinear set of data differs from the linear representation.

The items in the section were selected to provide an example of each of the expectations in the Algebra Standard. In addition, these assessment items were intentionally selected for several unique features. They include a range of tasks from informal to more formal algebraic approaches to solving problems. The set presents examples of assessing for students' understanding of algebra concepts and skills across the middle school grades, not just at the end of eighth grade. A heavy emphasis is placed on (a) representation; (b) moving among models, graphs, tables of values, and equations; and (c) demonstrating understanding of such concepts as slope, intercept, and equality as they are represented in different forms.

Although this chapter presents a range of examples, it does not constitute a full assessment of students' understanding of these concepts nor their fluent application of the related skills. Students need multiple opportunities over time and across contexts to demonstrate fluency, flexibility, and a deep understanding of algebra concepts and skills.

Algebra Assessment Items

Standard: Understand patterns, relations, and functions

Expectation: Represent, analyze, and generalize a variety of patterns with tables, graphs, words, and, when possible, symbolic rules

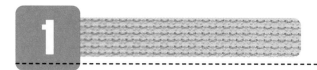

Dots can be arranged to form diamonds. Here are the first three diamond patterns.

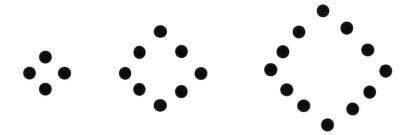

a. Make a drawing of the next diamond pattern.
b. How many dots will be in the sixth diamond pattern?
c. Complete the table.

Diamond Number	Number of Dots
1	4
2	8
3	12
4	
5	
6	

d.   Could a diamond pattern have 70 dots? Explain why or why not.

e.   How many dots would be in the *n*th diamond pattern? Justify your answer.

Source: Adapted from *Balanced Assessment Patterns and Symbols,* Mathematics in Context (New York: Holt, Rinehart &Winston, © 2000 by Encyclopædia Britannica)

About the mathematics: Students identify a missing element in a pattern, then apply the pattern to determine whether a specific solution will work given the pattern.

Solution

a.   A correct drawing has 5 dots on each side, and its adjacent sides share dots at the corners. The fourth diamond has 16 dots.

b.   24 dots

c.   The completed table is as follows:

Diamond Number	Number of Dots
1	4
2	8
3	12
4	16
5	20
6	24

d.   No, 70 is not a multiple of 4.

e. $d = 4n$ or $4n$. The pattern reveals that the numbers of dots are the multiples of 4.

Rubric (7 possible points)

a)   1 point for correctly drawn pattern

b)   1 point for the correct number of dots for the sixth term

c)   1 point for completing the table

d)   2 points for a correct answer that is justified appropriately

e)   2 points for a correct equation or expression with a justification of that equation or expression

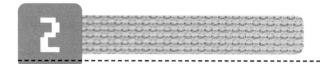

One apple costs $0.55, and one orange costs $0.75. The table represents the cost (in cents) of purchases of combinations of these items; for example, 1 apple and 3 oranges cost 280 cents.

Cost (in Cents) of Combinations of Apples and Oranges

	0	1	2	3	4	5	6	7	8	9	10
10	750	805	860	915	970	1025	1080	1135	1190	1245	1300
9	675	730	785	840	895	950	1005	1060	1115	1170	1225
8	600	655	710	765	820	875	930	985	1040	1095	1150
7	525	580	635	690	745	800	855	910	965	1020	1075
6	450	505	560	615	670	725	780	835	890	945	1000
5	375	430	485	540	595	650	705	760	815	870	925
4	300	355	410	465	520	575	630	685	740	795	850
3	225	280	335	390	445	500	555	610	665	720	775
2	150	205	260	315	370	425	480	535	590	645	700
1	75	130	185	240	295	350	405	460	515	570	625
0	0	55	110	165	220	275	330	385	440	495	550

Number of Oranges (vertical axis) / **Number of Apples** (horizontal axis)

a. Explain why the values in the table increase by 130 cents as one reads the numbers along a diagonal from the bottom left to the upper right.

b. Explain why the values in the table decrease by 20 cents as one reads the numbers along a diagonal from the upper left to the lower right.

c. If you had exactly $8.00 to spend on apples and oranges, what combination(s) of apples and oranges could you buy?

d. Write a formula that can be used to determine the cost of any combination of apples and oranges, letting C represent the total cost, A represent the number of apples, and O represent the number of oranges.

Source: Adapted from *Comparing Quantities,* Mathematics in Context (New York: Holt, Rinehart &Winston, © 1998 by Encyclopædia Britannica, T.E. p. 20)

About the mathematics: This item requires that students reason from a contextual problem presented in table form.

Solutions

    a. The values increase by 130 cents because each diagonal step to the right represents an increase of 1 apple and 1 orange. The cost of 1 apple at 55¢ and 1 orange at 75¢ is equal to 130¢.

    b. For each move from upper left to lower right, the values decrease by 20 cents because that amount is the difference between the cost of an orange and that of an apple: $75 - 55 = 20$.

    c. You could buy 5 apples and 7 oranges.

    d. $C = 55A + 75O$, where C represents the total cost, A represents the number of apples, and O represents the number of oranges.

Rubric (4 possible points)

1 point for each correct answer

0 points for no response or an incorrect answer

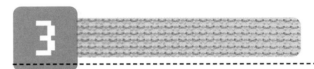

a. Find the 5th and 10th terms in the arithmetic sequence in table below.

Position in Sequence	1	2	3	4	5	...	10	n
Term	5	9	13	17	?	...	?	?

b. Using words or symbols, write a rule for the nth term in the arithmetic sequence.

Source: Adapted from Texas Education Agency (2003, released item)

About the mathematics: This item requires students to extend and generalize a sequence.

Solution

    a. The 5th term is 21; the 10th term is 41.

    b. The nth term is $4x + 1$, where x is the position in the sequence, that is, 4 times the position in the sequence plus 1 is equal to the number.

Rubric (4 possible points)

1 point for giving the 5th term in sequence

1 point for giving the 10th term in the sequence

1 point for work indicating that the sequence was extended past the 10th term

1 point for a generalization expressed either in words or with symbols

Teacher note: The option for expressing the rule for the *n*th term in words or symbols was intentional. Students should be encouraged to communicate with symbols and written rules.

4

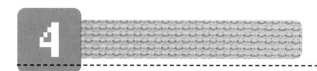

From a letter in the diagram at the right, a move can be made only to a letter diagonally adjacent and below. Following that pattern, how many different paths spell MATH? How do you know? Show your work.

About the mathematics: Here, students identify a pattern based on Pascal's triangle.

Solution: Eight such paths are possible. Breaking this pattern into smaller words reveals the number of paths that exist for words with various numbers of letters.

For the word MA, it is possible to see 2 paths.

2 paths

For the word MAT, it is possible to see 4 paths.

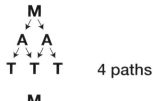

4 paths

For the word MATH, it is possible to see 8 paths.

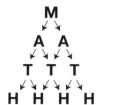

8 paths

Teacher note: This pattern lets students examine exponential growth. Many eighth-grade students in the pilot group were able to write an equation for this pattern as $p = 2^{r-1}$, where p is the number of paths and r is the number of rows.

Standard: Understand patterns, relations, and functions

Expectation: Relate and compare different forms of representation for a relationship

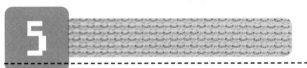

Ajax Taxicab Company charges a flat fee of $1.00 plus $0.30 per mile to ride in a cab. (Assumption: The flat-rate fee is incurred as soon as you enter the cab.) Do the relationships in the equation, graph, and table represent the relationship between the cost of riding in an Ajax Taxicab and the distance traveled? Explain why or why not.

Equation: $y = 0.03x + 1.00$, where y is the total cost of a cab ride and x is the distance traveled.

Price of a Taxi Cab Ride

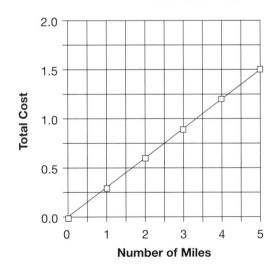

Price of a Taxi Cab Ride

Number of Miles (x)	Cost (y)
0	$1.00
1	$1.30
2	$1.60
3	$1.90
4	$2.20
5	$2.50

Source: Adapted from New England Common Assessment Program Grade-Level Expectations Resource Materials (Concord, N.H.: New Hampshire Department of Education, Rhode Island Department of Education, and Vermont Department of Education, October 2004)

About the mathematics: To be successful with this item, students need to understand that for any relation, three unique representations of the same information are possible.

Solution: The three representations do not represent the same relation or function. The equation and the table represent the relationship in the problem; the graph, however, does not. The situation is a nonproportional linear relationship, not proportional as the graph implies. The graph does not consider the "flat fee" of $1.00.

Standard: Understand patterns, relations, and functions

Expectation: Identify functions as linear or nonlinear and contrast their properties from tables, graphs, or equations

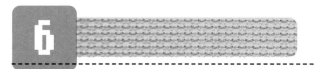

a. Which table or tables show the patterns of a linear relationship?

b. Describe how you decided whether the relationship between the variables in each table was linear.

Table A

x	y
–3	3
–2	2
–1	1
0	0
1	1
2	2
3	3

Table B

x	y
–3	–7
–2	–5
–1	–3
0	–1
1	1
2	3
3	5

Table C

x	y
–3	1
–2	$1\frac{2}{3}$
–1	$2\frac{1}{3}$
0	3
1	$3\frac{2}{3}$
2	$4\frac{1}{3}$
3	5

Source: Connected Mathematic Project Computer Test Bank (Upper Saddle River, N.J.: Pearson Education, 2000). CD-ROM.
About the mathematics: This item assesses students' ability to identify linear relationships among data presented in a table.

Solution

a. Tables B and C are both linear functions.

b. In table B, the values of y increase by 2 for every increase of 1 in x.
   In table C, the values of y increase by 2/3 for every increase in 1 of x.
   In table A, the y-values decrease by 1 and then increase again.
   A constant rate of increase in the dependent variable signals a linear relationship.

Standard: Represent and analyze mathematical situations and structures using algebraic symbols

Expectation: Develop an initial conceptual understanding of different uses of variables

---

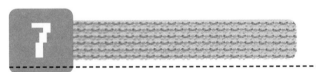

The rule for this table is $y = 3x + 4$. What number should replace the question mark in the y-column?

x	y
2	10
0	?
−1	1
−2	−2

a. $-\frac{4}{3}$
b. 5
c. 4
d. 7

Explain your answer.

Source: Maine Education Assessment (2002, item 31). http://mainegovimages.informe.org/education/mea/2002ReleasedItems/ Mar02G4MathScoreResponses.pdf (accessed March 22, 2005).
About the mathematics: Students are required to substitute a value found in a table into an equation, and apply the order of operations.

Solution: c

Distractors

a. The student substitutes 0 for y, not x, and solves for the value of x:
$$y = 3x + 4; 0 = 3x + 4; -4 = 3x; x = -4/3$$

b. The student interprets 5 as being halfway between 1 and 10.

c. The correct response is
$$y = 3x + 4: y = 3(0) + 4, y = 0 + 4, y = 4.$$

d. The student substitutes correctly but multiples incorrectly, or substitutes 1 instead of 0 for the value of x.
$$y = 3x + 4: y = 3(0) + 4, y = 3 + 4, y = 7$$
$$y = 3x + 4: y = 3(1) + 4, y = 3 + 4, y = 7$$

Teacher note: This question assesses a student's ability to apply formal algebraic understanding when finding the value for y if $x = 0$ given $y = 3x + 4$ and a table of some values.

About the student work: Two examples in this set of student responses have a correct solution (student response A and student response B), and one is incorrect (student response C). All three responses exemplify the instructional advantage of asking students to include a rationale for the answer they select in response to a multiple-choice question.

Student Work

Student Response A

This solution is the result of the student's correctly substituting 0 for x in the equation $y = 3x + 4$ and solving the equation for y.

$$y = 3x + 4$$
$$y = 3 \cdot 0 + 4$$
$$y = 0 + 4$$
$$y = 4$$

Student Response B

This solution implies the correct substitution of 0 in the equation.

Zero multipyed by any number, is still zero.
But zero plus 4 would be 4.

Student Response C

In this incorrect response, the student has ignored the equation. The solution is incorrectly based on the rationale that because 0 is halfway between 1 and –1, then 5 should be halfway between 10 and 1.

because it went "2" to "0" to -1
So it had to be "10", "5" to "1".

Standard: Represent and analyze mathematical situations and structures using algebraic symbols

Expectation: Explore relationships between symbolic expressions and graphs of lines, paying particular attention to the meaning of intercept and slope

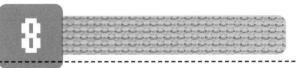

Which of the graphs on the following page best represents the graph of $y = x + 3$?

F. **G.**

H. **J.**

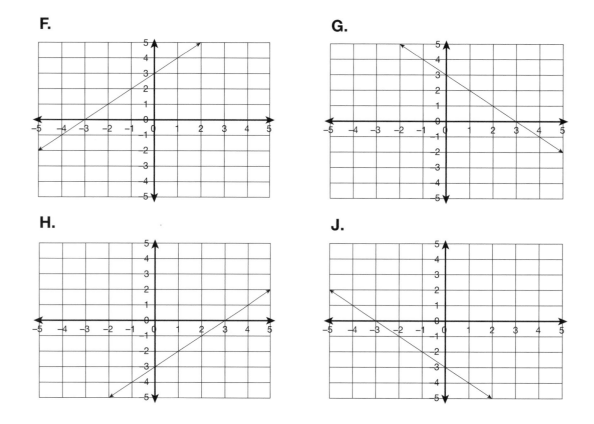

Source: Adapted from Virginia Standards of Learning Test (2002).
http://www.pen.k12.va.us/VDOE/Assessment/Release2002/Grade8/
VirgOnLine_Gr8_Math_1-23.pdf, item 58 (accessed March 22, 2005).

About the mathematics: This question assesses the student's understanding of the relationship between an equation and the graph of the equation.

Solution: F is the correct response.

Distractors

F. Graph F is the correct answer (the *y*-intercept is 3, and the slope is 1).

G. In this graph, the *y*-intercept is 3 but the slope is –1.

H. In this graph, the *y*-intercept is –3 with a slope of 1.

J. Here, the *y*-intercept is –3 with a slope of –1.

Teacher note: This question assesses the student's understanding of the relationship between an equation and the graph of the equation. It can be solved by identifying the slope (+1) and *y*-intercept (+3) in the given equation, then finding the graphical representation that matches the slope and intercept.

About the student work: The two responses included in this set illustrate students' confusion between the *x*- and *y*-intercepts. In both instances the student selected her or his answer on the basis of a positive value for the slope.

Student Work

Student Response A

In this response, H is incorrectly selected on the basis of 3 as the *x*-intercept and on the basis of a positive value for the slope.

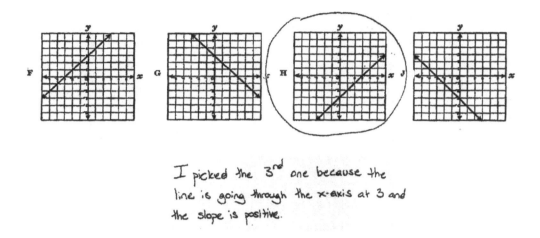

I picked the 3rd one because the
line is going through the x-axis at 3 and
the slope is positive.

Student Response B

In this response the correct distractor was selected on the basis of a positive value for the slope, but the evidence in the explanation illustrates the student's confusion about the *x*- and *y*-intercepts. This student's solution illustrates the importance of asking for a rationale for a student's choice of answer when using multiple-choice questions for classroom assessment.

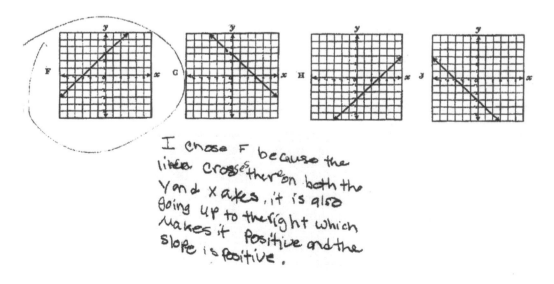

I chose F because the
lines crosses there on both the
Y and X axes, it is also
going up to the right which
makes it Positive and the
slope is Positive.

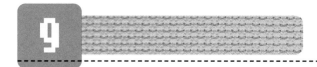

Francisco was working on a problem on his graphing calculator. He saw the following two screens:

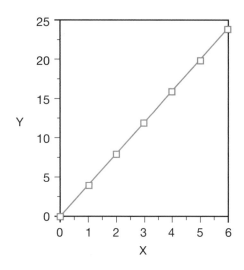

X	Y_1
0	0
1	4
2	8
3	12
4	16
5	20
6	24

a. Using words or symbols, write a rule that could be used to re-create the data in this table and graph.
b. Listed below are three relationships. Explain why your rule can or cannot be applied to each relationship.
    1. The relationship between the length of the side of any-sized square and its area
    2. The relationship between the length of the side of any-sized square and its perimeter
    3. The relationship between the width of any-sized rectangle and its perimeter

About the mathematics: This item involves writing a rule representing information provided in a graph and table, and reasoning about applications of the rule.

Solution

a. $y = 4x$.

b. 1. The area of a square is the square of its side and would be a nonlinear equation, so the rule cannot be applied.

   2. The perimeter of a square with sides of any length would be 4*s*, so the rule can be applied in this instance.

   3. The formula for the perimeter of a rectangle is $p = 2 \times 1 + 2 \times w$. In this situation we are concerned with only one variable; hence the rule will not apply.

Rubric (4 possible points): 1 point for each correct answer

About the student work: The two responses included in this set illustrate an understanding of the relationship represented in the graph and table. However, student A relates the relationship to the formula for area and perimeter, whereas student B gives an example in question b1 and states a generalized relationship (that the length of the side of a square is 1/4 its perimeter) in question b2. Of interest in the responses to question b3 is that student A erroneously states that the lengths could not be the same for a rectangle, whereas student B indicates that the relationship is not true in all cases. The formulas for both responses are correct.

Student Work

Student Response A

Responses to questions b1 and b2 are correct and supported with the formulas for the area and perimeter of a square. The response to question b3, however, has a major flaw. The student erroneously states that the lengths of the sides of a rectangle cannot be the same. The equation is correct.

Part a

$X \cdot 4 = Y$

Part b1

no the rule wouldn't work because if the length of each square = X than X •4 sides would equal the perimeter. Not the area. The area = BxH.

Part b2

yes the rule would work because,
if the length of each square = x than x·4
sides would equal the perimeter or (y)

Part b3

no the rule wouldnt work because all
of the lengths coddnt coldnt be the
same size. x·4 would work for a square, not
a rectangle

Student Response B

All three responses are correct. Question b1 is supported by only a single example. Question b2 is supported by a general statement about the relationship between the length of a side and the perimeter ("…the length of a side of square is 1/4 of its perimeter"). Question b3 shows an understanding that a square is a type of rectangle, and that the relationship would not hold up for rectangles that are not squares. The equation is correct.

Part a

$$y = x \cdot 4$$

Part b1

Cannot be, if a square has a side length
of 2, its area would be 4.

Part b2

Can be the length of a side of a square
is ¼ of its perimiter.

Part b3

Can be but not neccisary, a rectangle could be
45 inches long and 1 inch wide and that
would not be true.

Standard: Represent and analyze mathematical situations and structures using algebraic symbols

Expectation: Recognize and generate equivalent forms for simple algebraic expressions and solve linear equations

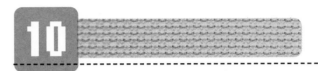

You are given the costs to purchase different amounts of candy.

Cost $9.75

Cost $9.50

a. What is the price of one lollipop?
b. What is the price of one chocolate bar?

Explain your answer

Source: Adapted from *Comparing Quantities,* Mathematics in Context (New York: Holt, Rinehart &Winston, © 1998 by Encyclopædia Britannica, p. 16)

About the mathematics: The student is asked to find solutions to a system of linear equations.

Solution

Each lollipop has a value of $2.50, and each candy bar has a value of $2.25.

Rubric (3 possible points)

3 points: Supplies correct values for both the candy bar and the lollipop, with supporting evidence that includes an approach that makes sense for the mathematics in the problem, and arrives at a solution set that works for both equations

2 points: Gives evidence of an approach that would work for determining the correct solution set if errors in the solution had not led to the wrong solution set

1 point: Demonstrates an understanding of equivalence in some aspect of the problem, even though the reasoning could not lead to the correct solution, for example, supplies solutions that would work for only one of the equations, not both

0 points: Attains incorrect solutions supported by inappropriate strategies

Teacher note: This question is included because it helps build mathematical understanding of systems of linear equations through an informal problem-solving situation.

About the student work: The strategies evidenced during the pilot test ranged from guess and check to the use of the substitution method for solving linear systems. The examples shown include a solution using the substitution method (student response A), a solution using guess and check (student response B), and a solution (student response C) evidencing major misunderstanding.

Student Work

Student Response A

In this solution the models (candy bar + lollipops) are translated into two equations, $C + 2L = \$9.75$ and $2C + 2L = \$9.50$. The variables are not defined in the response but are implied by the correct translation of the models into equations. The solution set for the two equations is determined using the substitution method of solving systems of linear equations.

Part a $2.50

Part b $2.25 $\begin{cases} c+3L = 9.75 \\ 2c+2L = 9.50 \end{cases}$

Part c

$$-2 \cdot (c + 3L = 9.75)$$
$$2c + 2L = 9.50$$
$$-6L = -19.5$$
$$2L = 9.50$$
$$-4L = -10$$

Lollypop = $2.50

$$c + 7.5 = 9.75$$
$$-7.5 \quad -7.5$$

Candy bar = $2.25

Student Response B

In this solution the student estimates that the values for the candy bar and lollipops are less than $3.00. The correct solution set is arrived at through a guess-and-check approach.

Part a $2.50

Part b $2.25

Part c

I guess kept checking until
I found the right answer. I knew that
The cost of the lallypop and the chocolate were
going to be less than 3 and I just punched numbers
in like 2.5 and 2.25 and added and got my
answer.

Student Response C

In this response the initial method of determining the value of a lollipop is incorrect ($9.75/4 = $2.43 rounded to $2.50). The value of the lollipop is then used in the first equation to find the value of a candy bar. Because $2.50 is the correct cost of a lollipop, then the correct cost for a candy bar is determined. The student seems to understand that the solution set must work for both equations, as evidenced by substitution into the second equation.

Part a   2.50

Part b   2.25

Part c

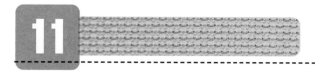

1) $9.75 \div 4 = 2.43$ round it to $2.50

2) $2.50 \times 3 = $7.50

3) 9.75
    -7.50
    $$2.25$

4) $2.25 + 2.25 + 2.50 + 2.50 = $9.50

11

During Rachel's birthday party, Rachel thinks about the ages of her parents and herself. Rachel says, "Hey, Mom and Dad, together your ages add up to 100 years!" Her dad is surprised. "You are right," he says, " and your age and mine total 64 years." Rachel replies, "And my age and Mom's total 58." How old are Mom, Dad, and Rachel? Explain how you got your answer.

Source: Adapted from *Comparing Quantities*, Mathematics in Context (New York: Holt, Rinehart & Winston, © 1998 by Encyclopædia Britannica, T.E. p. 82)
About the mathematics: This item requires students to use logical reasoning and number sense.
Solution

Mom	Dad	Rachel	Total Age
1	1		100
1		1	58
	1	1	64
1	1	2	122
		2	22
		1	11
1			47
	1		53

Alternatively, where M represents the mother's age, D represents the father's age, and R represents Rachel's age, then

$$M + D = 100,$$
$$M + R = 58,$$
$$D + R = 64.$$

Adding equations 2 and 3 yields $M + D + 2R = 122$. Subtracting equation 1 yields $2R = 22$. Therefore, $R = 11$, $M = 58 - 11 = 47$, and $D = 64 - 11 = 53$.

Standard: Use mathematical models to represent and understand quantitative relationships

Expectation: Model and solve contextualized problems using various representations, such as graphs, tables, and equations

12

The graph below shows Carlos's speed on his trip to school.

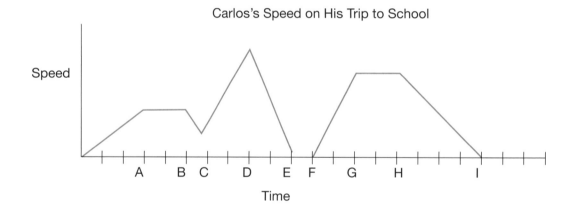

Carlos's Speed on His Trip to School

On the basis of the graph, when is Carlos's speed decreasing the most rapidly?

a. For the times between B and C
b. For the times between D and E
c. For the times between E and F
d. For the times between F and G

Explain your answer.

Source: Adapted from Massachusetts Comprehensive Assessment System (MCAS): Grade 8 (Malden, MA: Massachusetts Department of Education, 2003). www.doe.mass.edu/mcas (accessed March 22, 2005).

About the mathematics: Students interpret a nonnumerical graph, and interpret change on the basis of the slope at different time intervals.

Solution: b

Distractors

a. In this interval, Carlos's speed is decreasing, but not at as great a rate as in time interval DE.

b. This solution is correct.

c. The absence of positive or negative slope is confused with no motion of the vehicle.

d. This answer indicates confusion between increasing speed and decreasing speed.

Teacher note: This question assessing a students' informal understanding of slope, change, and the relationship between dependent and independent variables.

About the student work: In general, students in the pilot test selected the correct response for this question. The explanations usually referenced the steepness of the line, but only a few included discussions about negative slope as it relates to the context. Many of the responses included a reference to the steepness of the line, and some made a direct reference to the slope. The three examples provided in this set were selected because they illustrate (1) the importance of not relying on a single assessment question to draw conclusions about what students do or do not know; (2) the instructional advantage at the classroom level of asking students their rationale for selecting an answer when giving multiple-choice questions; and (3) the importance of asking follow-up or probing questions to pinpoint potential misunderstandings.

Student Work

Student Response A

This incorrect response illustrates the importance of asking students for a rationale for their choice of answer. Although this response shows evidence of an understanding that decreasing speed is related to downward slope, it also reveals a potential misconception that not "hitting the bottom" is important.

Answer a

It is between B and C because that is when the graph slopes down the most without hitting the bottom.

Student Response B

This response is correct. It includes a reference to slope, to the direction of the slope, and to the magnitude of the slope. It also points out the potential of using this type of assessment item to help students clarify their use of mathematical language.

Answer b

It decreases the fastest as you can see by looking at it. It has the biggest slope, almost straight down.

Student Response C

Although this student selected the correct answer from the distractors given, the rationale for that selection is not strong. Referencing change "in 1 letter" is not the same as referencing the distance traveled. Asking additional questions, for example, "What do you mean by 'in 1 letter?'" or "Interval pairs B–C and H–I both show decreasing speeds in the 1-letter interval; how are they similar to, or different from, interval E–F?" can help clarify the student's reasoning.

Answer b

Because it went from the highest point to the lowest point in 1 letter

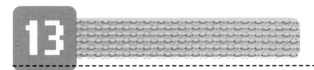

13

One of the most common codes for writing secret messages involves assigning a numerical value to each letter in the alphabet, for example,

$$A = 1, B = 2, C = 3,$$

and so on. (Each letter can have only one value assigned to it, and the numbers can be only positive and negative whole numbers.)

a   What numbers are assigned to O, N, and E in this coding system?
b. Add the numbers for O + N + E. What is the value of ONE? It would be nice if O + N + E = 1.
c. Assign integers to the letters so that O + N + E = 1 and T + W + O = 2.
d. With your assigned values, would T + E + N = 10?
e. Assign integers to the letters so that ONE = 1, TWO = 2, and TEN = 10.

> **Source:** Adapted from *Operations*, Mathematics in Context (New York: Holt, Rinehart & Winston, © 1998 by Encyclopædia Britannica, T.E. p. 144)
> **About the mathematics:** This item fosters reasoning with signed numbers.
> **Solution**
>   a. E = 5, N = 14, and O = 15.
>   b. ONE = 34.
>   c. Answers will vary. For example, if E = 1, N = 6, O = –6, T = 5, and W = 3, then O + N + E = 1 and T + W + O = 2.
>   d. No, T + E + N = 12. (The answer could also be "yes" depending on the values assigned in part c.)
>   e. Answers will vary. For example, if E = –1, N = 8, O = –6, T = 3 , and W = 5, then ONE = (–6) + 8 + (–1) = 1, TWO = 3 + 5 + (–6) = 2, and TEN = 3 + (–1) + 8 = 10.

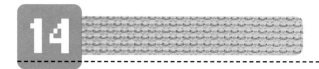

The graph below shows Carlos' speed on his trip to school.

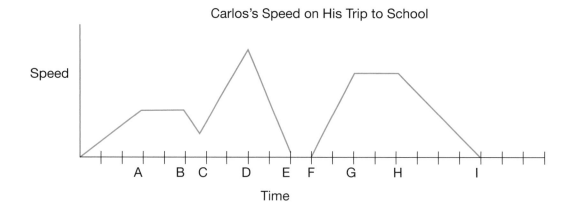

On the basis of the graph, for which of the following times, other than before he starts, is Carlos stopped?

a. For the times between A and B
b. For the times between E and F
c. At time D
d. For the times between B and C

Explain your answer.

> **Source:** Adapted from Massachusetts Comprehensive Assessment System (MCAS): Grade 8 (Malden, MA: Massachusetts Department of Education, 2003, p. 175, item 12). www.doe.mass.edu/mcas (accessed March 22, 2005).
> **About the mathematics:** Students interpret a nonnumerical graph, and interpret change on the basis of the slope at different time intervals.
> **Solution:** b

Distractors

    a. This answer indicates confusion between a steady rate and no motion of the vehicle.

    b. This answer is correct.

    c. This response indicates confusion between the top speed, or the point at which motion changes, with no motion of the vehicle.

    d. This answer indicates confusion between decreasing speed and no motion of the vehicle.

Teacher note: Teachers could ask many more questions using this scenario as a prompt, for example, "During which interval was Carlos traveling at the greatest speed for the longest period of time?" or "During which interval might Carlos have been biking up a hill? Why?"

About the student work: In general students in the pilot test selected the correct response and gave strong rationales for their selection. The most common error that students made was selecting the interval A–B, as in the single student response included here.

Student Work

Student Response

The incorrect interval between A and B was selected. The rationale indicates confusion between a straight line and no speed.

> I picked this one because the line is straight between A+B so it means he's not doing any work.

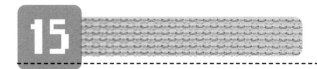

Use the balance scale below to answer the following question.

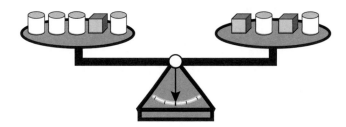

Which of the following shows the relationship between the weight of one cylinder and the weight of one cube?

a. One cube weighs the same as two cylinders.
b. One cube weighs the same as four cylinders.
c. One cylinder weighs the same as two cubes.
d. One cylinder weighs the same as four cubes.

Explain your answer.

> **Source:** Adapted from Massachusetts Comprehensive Assessment System (MCAS): Grade 8 (Malden, MA: Massachusetts Department of Education, 2001, p. 299, item 15). www.doe.mass.edu/mcas (accessed March 22, 2005).
> **About the mathematics:** This item involves determining the relationship (linear) between two unknown values of an equation from a model.
> **Solution:** a (1 cube = 2 cylinders)
> **Distractors**
> a. This answer is correct.
> b. This response indicates the use of the relationship on the left side of the balance, not the relationship between the two sides of the balance.
> c. This response indicates reversal of the values for the cube and the cylinder.
> d. This answer indicates that the cylinder is erroneously seen as being heavier than the cube.
> **Teacher note:** This question is an example of assessing informal algebraic thinking.

About the student work: In general, students in the pilot test selected the correct response. The most common error is exemplified in student A's work. This response shows evidence of understanding that the cube is heavier than the cylinder, but the correct relationship is not determined. Student B's response illustrates the importance of asking students for the rationale for their answer. In this instance the student arrived at the correct answer for the wrong reason, pointing out a flaw in the original question. We encourage teachers to ask that students give the rationale for their answer when using this assessment item. The different models or explanations in the responses of students C, D, and E show evidence of emerging understanding of the need to keep the equation balanced to determine the correct relationship between the number of cylinders and the number of cubes.

Student Work

Student Response A

In this incorrect response, the student appears to understand that the cube is heavier than the cylinder. However, the correct relationship between the cube and cylinder is not established.

Answer b

I picked (B) Because since a cube is heavier than a cylinder itt needs more cylinders than 1 cylinder. 1 cube and 4 cylinders!

Student Response B

In this response the correct distractor is selected for the wrong reason. The ratio of the total number of cylinders to the total number of cubes is used. This answer presents a good opportunity to ask, "Will this approach work if a cylinder is added to each side of the balance? Why or why not?"

Answer a

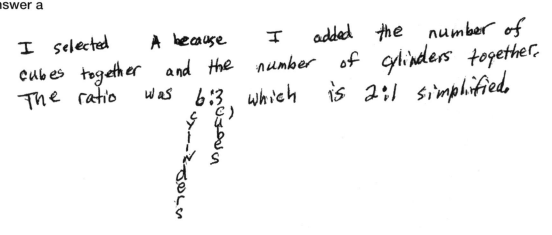

I selected A because I added the number of cubes together and the number of cylinders together. The ratio was 6:3 which is 2:1 simplified.

Student Responses C, D, and E

All three students selected the correct distractor from the choices. They also provided different types of models or explanations as evidence of their informal understanding of solving equations.

Response C

Answer a

After I crossed out 1 cube on each side There was this ➝ and after crossing out 2 on each side

I had this

Response D

Answer a

$$\cancel{4}\square \; | \square \;\; = \;\; \square\square \; \square\square$$

$$\times\times\times\cancel{4}\cancel{\square} \;\; = \;\; \square b \cancel{4}\times \qquad \square \cancel{\square}\cancel{\square} = \cancel{4}\cancel{4}\times\times\cancel{\square}$$

$$\times\times = \square$$

Response E

Answer a

cross out on each side what they have in common, such as they both have a cube. If you take away 2 cube from each and do that for as many as possible it would still be equal because you took away the same amount of each. You are left with a cube on one side and 2 cylinders on the other meaning they are equal to eachother.

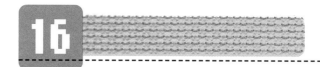

The Smith Family is planning to subscribe to an Internet service. They have three payment plans from which to choose, and have asked for your help in choosing which one to purchase.

The table below provides the rates for each of the plans.

	Monthly Rate	Hours of Web Access for the Monthly Rate	Cost for Each Additional Hour above the Monthly Rate
Plan A	$14.00	15	$2.00
Plan B	$10.00	10	$2.50
Plan C	$20.00	Unlimited Use	$0.00

Evaluate each plan, and then write a letter to the Smith family with–

- a recommendation for which plan to buy;
- the advantages and disadvantages of each plan; and
- the number of hours for which each plan would cost the least.

Provide specific costs in tables or graphs that support your recommendation, the advantages and disadvantages of each plan, and the number of hours for which each plan would cost the least.

Source: Adapted from *MathThematics—Book Three* (New York: Houghton Mifflin Co., © 1998 by McDougal Littell, p. 497)

About the mathematics: In this item, students compare costs of three plans on the basis of different rates.

Solution: Which plan is best depends on the number of hours accessing the Web during a one-month period. Plan C is the best plan for anyone accessing the Web more than 17 hours per month. Plan B is best for anyone accessing the Web up to 11 hours, and plan A is best for anyone accessing the Web between 11 and 17 hours. The shaded areas in the table indicate the numbers of usage hours for which each plan offers the lowest cost. The graph shows the relationship between the cost and number of hours used for the three plans.

Number of Hours	Cost (in Dollars) of Internet Access Plans		
	A	B	C
0	14	10	20
1	14	10	20
2	14	10	20
3	14	10	20
4	14	10	20
5	14	10	20
6	14	10	20
7	14	10	20
8	14	10	20
9	14	10	20
10	14	10	20
11	14	12.5	20
12	14	15	20
13	14	17.5	20
14	14	20	20
15	14	22.5	20
16	16	25	20
17	18	27.5	20
18	20	30	20
19	22	32.5	20
20	24	35	20
21	26	37.5	20
22	28	40	20
23	30	42.5	20
24	32	45	20

Accessing the Web

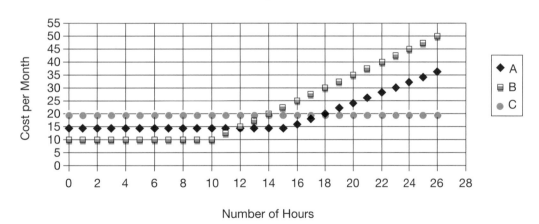

Teacher note: This problem involves applying rates to solve a problem with a familiar context.

About the student work: Both responses contain errors that present an instructional opportunity. The plans in response A are represented with different graphs having different scales, making any comparison across plans difficult. In response B the graph accurately reflects the situation; however, the interpretation was made at exact times instead of time intervals.

Student Work

Student Response A

In this response the correct costs at five-minute time intervals (and one-minute intervals) are determined for each plan. Although graphs are included, they all employ different scales and so cannot be used for direct comparisons of the plans. In addition, because the graphs are based on plotted points in five-minute intervals, the time interval between 0 and 5 minutes for each plan is incorrect. The rationale offered for the choice of the best plan is based on the five-minute time intervals. Note also that the graphs are titled, but except for plan A, the axes are not labeled.

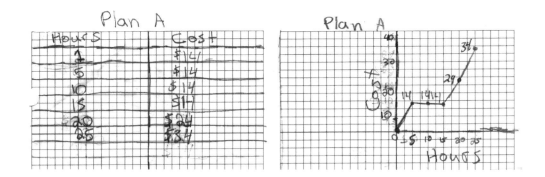

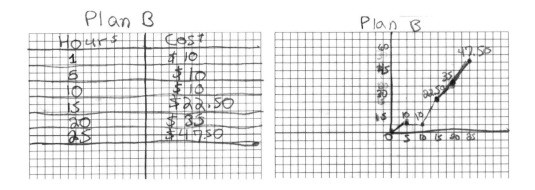

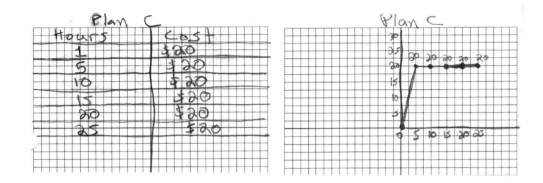

Dear Smiths,

After going over these three plans I noticed that Plan C was the best plan. For Plan A the price got higher for every hour you were on. It didn't grow as fast as plan B. Plan B is about $12 higher than A. If you don't get the internet much A or B would be the best plans to choose.

Plan C is great for if you go on internet a lot. For little internet use I would not use this plan. The least number of hours for A is 15. B is 10. C is 20.

Student Response B

This response includes an accurate graph and tables to represent the situation.
However, the interpretation of hours of use for which each plan would be best is
made for a single length of time, not a time interval.

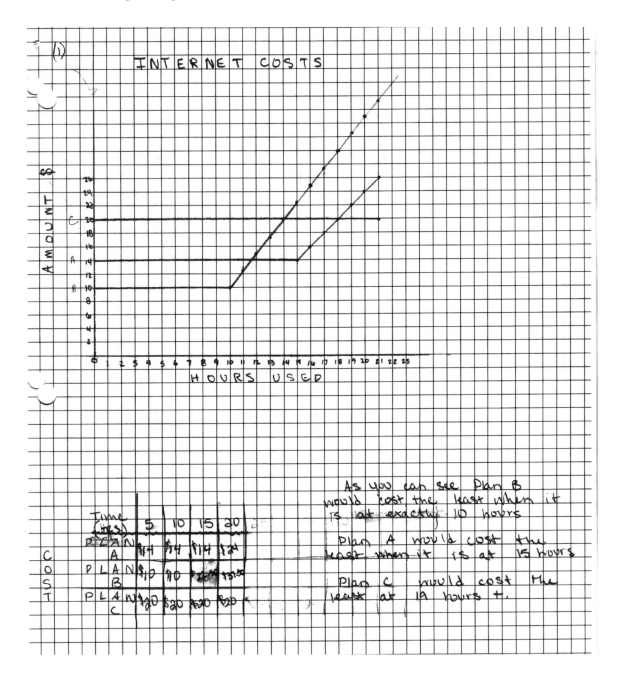

Standard: Analyze change in various contexts

Expectation: Use graphs to analyze the nature of changes in quantities in linear relationships

Standard: Use mathematical models to represent and understand quantitative relationships

Expectation: Model and solve contextualized problems using various representations, such as graphs, tables, and equations

17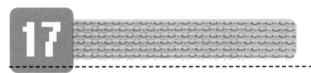

You work in a factory that makes shoes. You have to decide what length shoelace to select when you know the number of lace holes in a shoe. The shoes in your company have from 1 pair of lace holes up to 15 pairs of lace holes. A full-size shoe with 7 pairs of lace holes is shown [on the next page] to help you solve this problem.

1. Create a rule that will tell you the length of the shoelace needed when the number of holes in the shoe is known. Explain how you arrived at the rule.
2. Design three signs to display in the factory to help people select the correct shoelaces if they know the number of pairs of lace holes.
   a. One sign must have a table.
   b. One sign must have a graph.
   c. One sign must have a formula.

Source: Adapted from The Balanced Assessment Project (unpublished)

About the mathematics: Students are asked to represent a linear relationship in an equation, table, and a graph.

Solution: Rule for the length of shoelace needed when the number of lace holes is known:

> The approximate length of the shoelace needed to tie the shoe is 36 centimeters (or 14 in.) plus the approximate length needed to span the first pair of lace holes (5 cm, or 2 in.), or about 41 centimeters (16 in.); plus the approximate length needed between pairs of lace holes(10cm, or 4 in.).
>
> If t is the total length of shoelace, and x is the number of pairs of lace holes, then $t = 41 + 10x$ (in centimeters) or $t = 16 + 4x$ (in inches).

	Approximate Length	
	Centimeters	Inches
Approximate length needed to tie shoe	36	14
Approximate length (in cm) between first pair of holes	5	2
Approximate length (in cm) needed between pairs of holes	10	4

Number of Pairs of Holes	Length (in cm) Needed	Length (in Inches) Needed
1	36 + 5 = 41	16
2	51	20
3	61	24
4	71	28
5	81	32
6	91	36
7	101	40
8	111	44
9	121	48
10	131	52
11	141	56
12	151	60
13	161	64
14	171	68
15	181	72

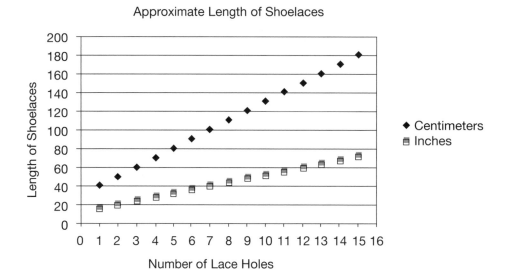

Approximate Length of Shoelaces

Rubric (8 possible points)

    1 point for a solution that accounts for length of shoelace to tie shoe ($\approx$14 in./36cm), and the distance across the first two holes ($\approx$2 in./5 cm)

    1 point for solution that accounts for an additional approximately 4 inches for each additional pair of holes

    2 points for a rule or equation that includes both the additional shoelace needed for each pair of holes (slope) and the length for tying the shoe plus the distance across the first pair of holes (intercept). Equations will vary depending on measurements but should reflect the following:

y = (inches needed between each pair of holes)x
        + (initial length between first pair of holes + lace needed to tie shoe)

(i.e., $y = 4x + 16$, where y represents total length needed on the basis of the number of pairs of holes, 4 represents the slope [increase per hole], and 16 is the initial amount of lace needed [intercept]).

    3 points for a graph that accurately and appropriately represents the situation in the table and rule (1 point for accurate scales and axis, 1 point for correct intercept, and 1 point for correct translation of the data)

    1 point for accurate organization of the data in the table

Teacher note: In this problem-solving situation, students work with linear relationships in a context. It is an excellent opportunity to assess the ability to represent a contextual situation with an equation, graph, and table.

About the student work: In general, the responses from the pilot test were mixed. The two responses included here reflect some of the difficulties that students experienced.

Student Work

Student Response A

This response includes conflicting information. The explanation indicates that the length of lace needed to tie the shoe is 12 inches, but the y-intercept used in the table, the equation ($y = 4x + 10$), and the graph is 10 inches. However, if one assumes that the y-intercept is 10, then the equation, table, and graph are accurate representations of the situation.

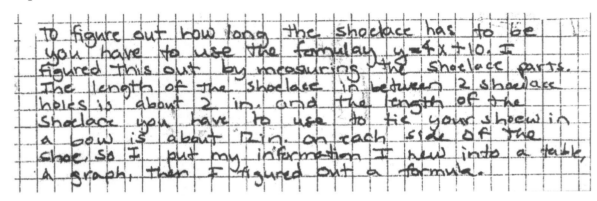

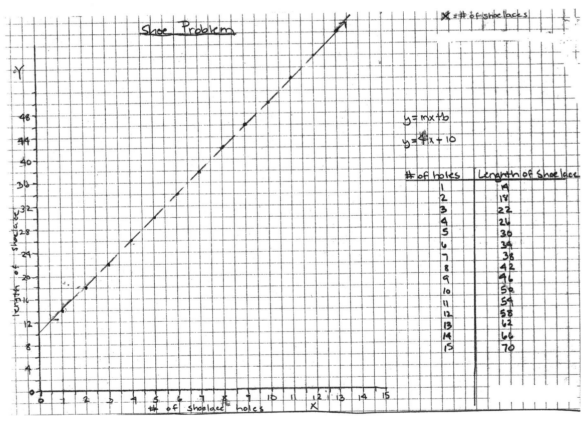

Student Response B

This response reflects an incorrect understanding of the situation. The total number of inches needed to tie a shoe with 7 pairs of lace holes is accurately determined through measurement (see page 82). However, the length needed to tie the shoe on the basis of each additional pair of lace holes has been determined by dividing the total (166 cm) by the number of holes (14). This solution does not account for the approximately 36 centimeters of shoelace needed to tie the shoe regardless of the number of lace holes, nor for the change of 10 centimeters per pair of lace holes.

Answer part 1

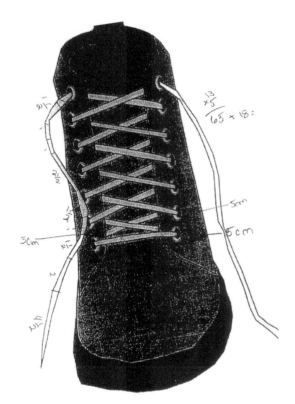

$166 \div 14 = 11.86$

You need 11.86 cm of lace per hole. I got that by first finding out about how long you need a shoe lace for a 14 hole shoe. Then I divided 14 holes into 166 cm and got 11.86

Answer part 2

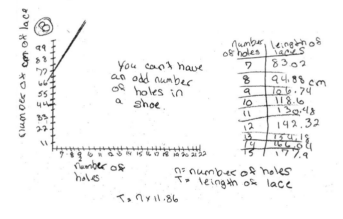

You can't have an odd number of holes in a shoe.

number of holes	leingth of laces
7	83.02
8	94.88 cm
9	106.74
10	118.6
11	130.48
12	142.32
13	154.18
14	166.04
15	177.9

n = number of holes
T = leingth of lace

$T = n \times 11.86$

18

Nanda has a **tall, thin** candle and a **short, thick** candle.

The tall, thin candle is 40 centimeters tall. It loses 3 centimeters in height for each hour it burns.
Here is a formula you can use to compute the height of the tall, thin candle after it burns for a given number of hours:

$$h = 40 \text{ centimeters} - 3x$$

(where x represents the number of hours that the candle burns and h represents the height of the candle).

a. What is the height of the tall, thin candle after it has burned for 4 hours? Explain you answer.

The **short, thick** candle is 15 cm tall. It loses a centimeter in height for each hour that it burns.

b. Create a formula to compute the height of the short, thick candle after it burns for a given number of hours.

c. Which of the two candles lasts longer: the **tall, thin** candle or the **short, thick** candle? Show your work.

d. Nanda thinks that if the **tall, thin** candle and the **short, thick** candle are lit at the same time and allowed to burn continuously, at one point in time they will be exactly the same height. Is Nanda correct?

If your answer is yes, tell when the two candles will be the same height.

If your answer is no, explain why the two candles will never be the same height.

Support your answer with *tables and graphs*.

Source: *Expressions and Formulas Balanced Assessment Unit,* Mathematics in Context (Holt, Rinehart & Winston, © 2000 by Encyclopædia Britannica, p. 3)

About the mathematics: This item requires students to represent a linear relationship with a negative slope in an equation, a table, and a graph.

Solution

a. 28 cm

b. Where h represents the height of the short, thick candle, and x represents the number of minutes the candle burns, then $h = 15 - x$.

c. On the basis of the graph and table, the short, thick candle lasts 1 minute longer than the tall, thin candle.

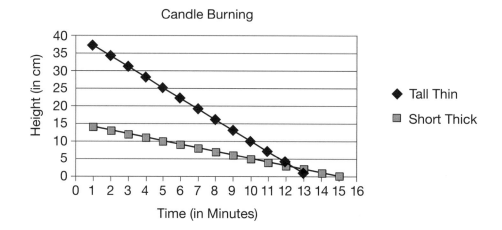

Hours	Tall, Thin	Short, Thick
0		
1	37	14
2	34	13
3	31	12
4	28	11
5	25	10
6	22	9
7	19	8
8	16	7
9	13	6
10	10	5
11	7	4
12	4	3
13	1	2
14		1
15		0

d. At 12.5 minutes they are the same height.

Hours	Height in cm	
	Tall, Thin	Short, Thick
0		
0.5	38.5	14.5
1	37	14
1.5	35.5	13.5
2	34	13
2.5	32.5	12.5
3	31	12
3.5	29.5	11.5
4	28	11
4.5	26.5	10.5
5	25	10
5.5	23.5	9.5
6	22	9
6.5	20.5	8.5
7	19	8
7.5	17.5	7.5
8	16	7
8.5	14.5	6.5
9	13	6
9.5	11.5	5.5
10	10	5
10.5	8.5	4.5
11	7	4
11.5	5.5	3.5
12	4	3
12.5	2.5	2.5
13	1	2
13.5	−0.5	1.5
14	−2	1

Rubric (10 possible points)

2 points for the answer of 28 cm as the height after 4 hours, supported with an explanation

2 points for correct formula with variables defined (15 cm – $1x = h$, where x represents the number of hours that the candle burns and h represents the height of the candle)

2 points for the response that the short, thick candle burns for 15 hours, whereas the tall, thin candle burns for 14 hours, with work that supports the answer

2 points for the answer that at 12.5 minutes the candles are the same height, with work that supports the answer

2 points for a graph that is accurate and appropriate, given the data

Teacher note: This problem is cast in a context with a negative slope.

About the student work: The four pieces of student work included here indicate the range of responses to this question. The most challenging parts of the problem are parts c and d. In some instances, such as student response A, an assumption is made on the basis of incomplete data. In other instances, such as student response C, the dependent and independent variables are switched when graphing the data, or the y-intercept is disregarded. Student response D is correct and is included by way of contrast with the other examples, which demonstrate errors.

Student Work

Student Response A

A calculation error is made in part a in determining the height of the tall candle after it burns for 4 hours ($3 \times 4 = 20$, not 12). However, that error is not reflected in other work associated with the tall candle (table associated with parts c and d). The formula for determining the height of the short, thick candle as is burns is accurate, although the variables are not explicitly defined. These understandings are reflected in the solution to parts c and d. However, the solutions to parts c (Which candle lasts longer?) and d (Are the candles the same height at any point if the candles start to burn at the same time?) are incorrect because an assumption is made on the basis of the first ten minutes of burning. Both the table and the graph stop at the end of ten minutes.

Part a

40 centimeters − 3×4 = 20 centimeters
tall thin candel after 4 hours = 20 centimeters

Part b

15 cm − 4 = 11

H = 15 cm − 1x = formula
to find rate of
burn per hour of time.

Part c

The tall thin candel

hour	1	2	3	4	5	6	7	8	9	10
tall thin	37	34	31	28	25	22	19	16	13	10
short thick	16	13	12	11	10	9	8	7	6	5

I found that the tall thin candel will last longer if both candels are lit at the same time.

Part d

Because they burn at different rates the short thick one will run out first.

they will never reach the same hieght at the same time if both are lit at the same time

15 cm = ——
40 cm = • • • •

Student Response B

This response includes accurate answers until part d. In part d the response indicates pairs of times at which the candles are the same height (e.g., at 11 and 8 hours they are, respectively, 7 cm high), not the single time that they are the same height. The response does not elucidate whether the question has been misunderstood, whether the solution to part c has not been considered in the student's thinking, or whether a graph of the situation would have clarified the student's understanding.

Part a

$$40 cm - 3.4 = H \quad 40 - 12 = 28 cm$$

Part b

$$H = 15 cm - 1x$$

Part c

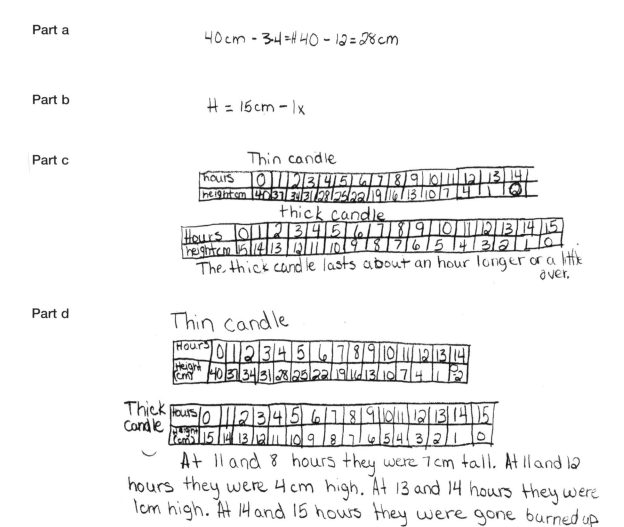

Part d

Student Response C

The graph included from this response exemplifies two major errors that were evidenced in some student work in the pilot test.

1) The dependent and independent variables are reversed. The height of the candle depends on the number of hours that the candle burns. Therefore, the number of hours should be graphed on the *x*-axis, and the height of the candle, on the *y*-axis.

2) The graph of the lines does not include the *x*-intercepts (given the reversal of the axis).

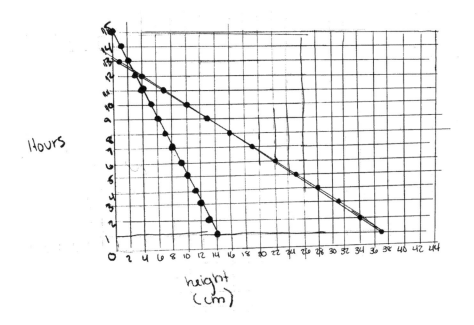

Student Response D

Part d from this response is included in contrast with the previously presented solutions to part d of the problem. In this response (see page 94) the table is used as a source of the data for the graph. The graph is then used to represent the data and determine the time at which the candles are the same height. The graph is properly labeled, titled, and scaled. In addition, the *y*-intercepts and slopes of each relationship are accurately graphed.

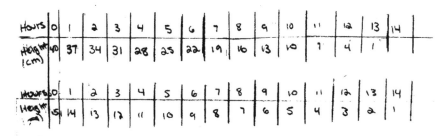

Hours	0	1	2	3	4	5	6	7	8	9	10	11	12	13	14
Height (cm)	40	37	34	31	28	25	22	19	16	13	10	7	4	1	

Hours	0	1	2	3	4	5	6	7	8	9	10	11	12	13	14
Height	15	14	13	12	11	10	9	8	7	6	5	4	3	2	1

Yes, because since you start off with different hieghts
of the candles and burn them for the same number
of hours the tall thin candle burn way faster
than the short thick one and eventually they
are at the same height at 12:30.
(See graph below)

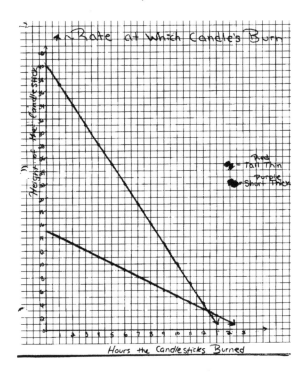

Rate at which Candle's Burn

Standard: Analyze change in various contexts

Expectation: Use graphs to analyze the nature of changes in quantities in linear relationships

Standard: Represent and analyze mathematical situations and structures using algebraic symbols

Expectation: Explore relationships between symbolic expressions and graphs of lines, paying particular attention to the meaning of intercept and slope

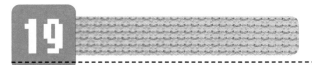

Tonya is draining all the water from a **full** aquarium to clean it. The graph below shows the amount of water left in the aquarium as Tonya drains the water.

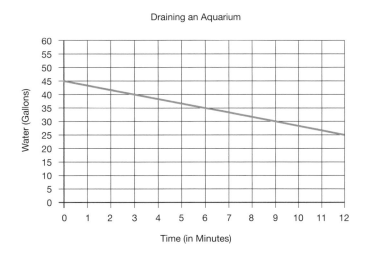

Draining an Aquarium

a. How much water was in the aquarium when it was full? Explain your reasoning.
b. How much water was drained from the aquarium in 1 minute? Explain your reasoning.
c. Write an equation that shows the amount of water (A) left in the aquarium after t minutes.
d. How many gallons are left in the aquarium after 10 minutes?
e. How much time will be needed to remove all the water from the aquarium? Explain your reasoning.

Source: Adapted from *Connected Mathematics Project* "Moving Straight Ahead" (Upper Saddle River, N.J.: Pearson Prentice Hall, 2002)

About the mathematics: This item requires students to explain the meaning of the intercept in the context, interpret a rate (slope) from the graph, write an equation to represent the situation, and apply the information (intercept and slope) on the graph to interpolate and extrapolate information.

Solution

a. 45 gallons, *y*-intercept

b. The siphon removes 20 gallons in twelve minutes, or $5/3$ gallons per minute.

c. $G = -5/3\,t + 45$, where G is the number of gallons in the tank and t is the time in minutes.

d. The number of gallons left after 10 minutes is $28^{1}/_{3}$ gallons ($G = -5/3$ (10 min.) + 45).

e. When 0 is substituted for G in the equation, then $t = 27$ minutes (0 gal. = $-5/3\,x + 45$).

Rubric (8 possible points)

a. 1 point for answer of 45 gallons; alternatively, 1 point for explanation that the number of gallons at the start is equal to the *y*-intercept (number of gallons at 0 minutes)

b. 2 points for the correct answer of $5/3$ gallons per minute, supported with reasoning consistent with the problem; alternatively, 1 point for estimating the rate from the graph

c. 2 points for the correct equation, or just 1 point if the slope is correct but the intercept is missing

d. 1 point for the correct number of gallons in the tank, that is, $28^{1}/_{3}$ gallons after 10 minutes, or for basing the value on the rate identified in part b

e. 2 points for the correct elapsed time of 27 minutes, supported with reasoning consistent with the problem, or for a value that is based on the rate identified in part b

Teacher note: This question is based on the meaning and application of concepts of slope and intercept. It lets teachers assess students' level of understanding of these concepts, as well as their ability to recognize and apply the associated skills in a problem situation.

About the student work: The responses to this question ranged from interpolating the rate on the basis of an estimate of the change between the start and 1 minute to calculating the slope as $-5/3$. In many responses, as in student response A, the rate (gallons/minute) was estimated by interpolating the value at 1 minute as 2 gallons per minute. Student response A illustrates the instructional opportunity that can arise if students estimate the slope. A teacher might ask students to make a graph using the same scales as the original graph and compare the slope of the line created when using a slope of 2 with the line in the original graph. One might contrast this error, which is not crucial when emptying a fish tank, with a situation in which such an error could make a big difference (e.g., transfer of fuel in flight).

Student Work

Student Response

This response gives the correct amount of water in the aquarium when it is full. The amount of water drained per minute is estimated (2 gallons per minute), not calculated ($5/3$ gallons per minute). Although not the correct answer, the equation provided is accurately based on the interpolated rate. The solutions for parts d and e are also based on the interpolated, rather than calculated, rate.

Part a
    45 gallons, because when the line starts on the graph is where she starts to empty it, and it is on 45.

Part b
    about two gallons.

Part c
    $A = 45 - (t \cdot 2)$

Part d
    about 27 gallons

Part e
    $45 - (t \cdot 2) = 0$
    $+45 \quad +45$
    $\dfrac{t \cdot 2}{2} = \dfrac{45}{2}$
    $t = 22.5$

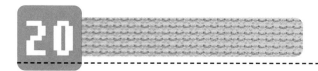

Different numbers of pamphlets are to be mailed using large boxes that, when empty, weigh 40 ounces each. Each pamphlet weighs 4 ounces. The empty space will be filled with packing pellets. (Since the pellets are very light, you can ignore the weight of the pellets.)

a.   Create a function that represents the relationship between the number of pamphlets in the box and the total mailing weight.
b.   Graph the function.

c.   Explain the role that the weight of the box and the weight of each pamphlet play in the graph in terms of slope and intercept.

> **Source:** Adapted from *Foundations of Success: Mathematics for the Middle Grades* (Washington, D.C.: Achieve, 2001, p. 129)
>
> **About the mathematics:** This item involves explaining the meaning of the intercept in context, and understanding the situation sufficiently well to write an equation that accurately represents the situation.
>
> **Solution**
> a.   The weight of the box is 40 ounces, and each pamphlet added weighs an additional 4 ounces. The equation that represents the total weight of the box and pamphlets is $w = 4x + 40$, where w represents the weight of the box and x represents the number of pamphlets.
> b.   The correct graph is shown below.

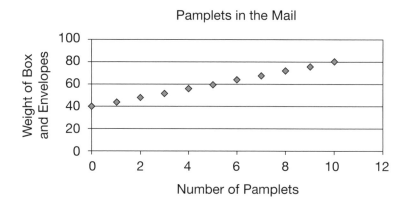

Pamplets in the Mail

    c. The slope is defined as the weight of each envelope (4 ounces), and the intercept, as the weight of the box (40 ounces).

Rubric (5 possible points)
    a. 2 points for the correct equation, expressed in either words or symbols, with variables defined; 1 point for the correct equation with variables undefined
    b. Graph: 3 points for correct dependent and independent variable; accurate scales; representation of intercept; and graph of the function plotted as points, not a line. Explanation: 2 points for correct explanation of both the slope and the intercept; 1 point for one or the other

Teacher note: This problem is included to assess some foundational understandings about representing a function, slope, and intercept in a concrete context.

About the student work: In general, students' responses included an appropriate expression or equation, in words or symbols, that represents the relationship. Although all responses included a graph of the equation, many revealed significant errors, such as passing the graph of the line through the origin, disregarding the original weight of the box (*y*-intercept), connecting the points, and reversing the dependent and independent variables.

Because this question is tightly tied to an understandable context, such errors can be turned into good instructional opportunities by asking such questions as the following: For graphs in which points are connected: "Would someone ship a fractional part of a pamphlet? Why or why not?" "What implication does this constraint have for constructing the graph of this function?" For graphs in which the line passes through the origin: "How does the weight of sending 10 pamphlets compare when the equation is used to determine total weight versus when the graph is used?" Why are the weights different?" For switching the axes: "Is the total weight of the box dependent on the number of pamphlets, or is the number of pamphlets dependent on the total weight of the box?"

Student Work

Student Response A

In this response the equation is accurately represented and the variables are correctly defined.

n = number of pamphlets
a = total weight of package (in ounces)
4n + 40 = a

Student Response B

In this response the equation is accurately represented in words.

mailing weight = number of pamphlet, times 4 ounces, plus 40 ounces.

Student Response C

In this response the dependent and independent variables are reversed (the total weight is dependent on the number of envelopes), and the points represented by the weight of each pamphlet are inappropriately connected by a line.

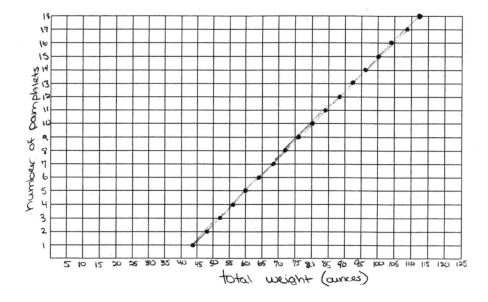

Student Response D

In this response the dependent and independent variables are correct, with the graph of the equation intercepting the *y*-axis at the weight of the box (*y*-intercept). However, the points represented by the weight of each pamphlet are inappropriately connected by a line, and a scaling error occurs on the *y*-axis.

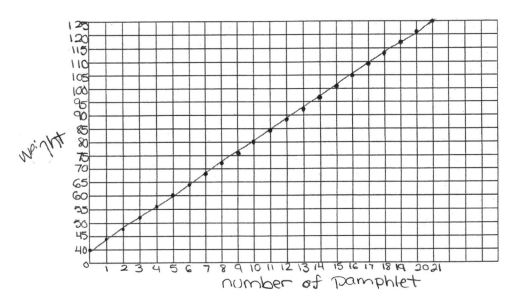

Student Response E

In this response the independent and dependent variables are reversed, and the graph of the function inappropriately passes through the origin, thus ignoring the initial weight of the box.

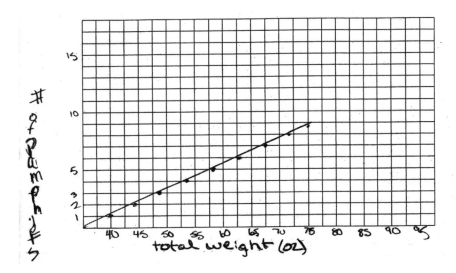

Student Response F

In this response the dependent and independent variables are correct, with the graph of the function intercepting the y-axis at the weight of the box (y-intercept). A ziz-zag line on the y-axis between 0 and 40 appropriately indicates that the starting scale is not the same as the other points on the scale. However, a line inappropriately connects the points represented by the weight of each pamphlet. In part c of the problem, the student seems to evidence some confusion about the meaning of the y-intercept ("The weight of the box and the weight of pamphlet makes the y-intercept…"). However, the meaning of the slope is accurately stated.

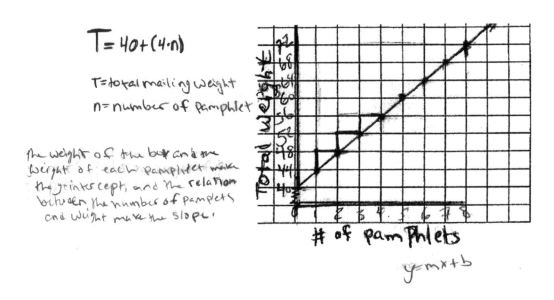

$$T = 40 + (4 \cdot n)$$

T = total mailing weight
n = number of pamphlet

the weight of the box and the weight of each pamphlet make the y-intercept, and the relation between the number of pamplets and weight make the slope.

# of pamphlets

$y = mx + b$

Standard: Analyze change in various contexts

Expectation: Use graphs to analyze the nature of change in quantities in linear relationships

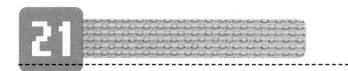

21

The three graphs below show the progress of a cyclist at different times during a ride. For each graph, describe the rider's progress over the time interval.

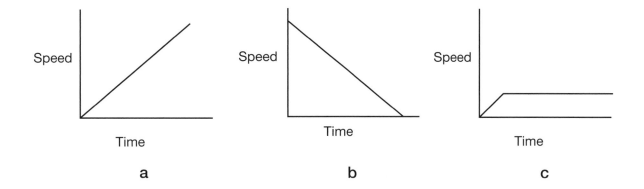

a   b   c

Source: *Connected Mathematics Project,* "Variables and Patterns" (Upper Saddle River, N.J.: Prentice Hall, 2000, p. 135)

About the mathematics: This item involves the interpretation of a nonnumerical graph.

Solution

    a. The speed is increasing over time.

    b. The speed is decreasing over time.

    c. The speed increases for a short amount of time, and then is constant.

Rubric (3 possible points): 1 point for each correct answer

Teacher note: This question assesses understanding of the meaning of the relationship between speed and time represented in a nonnumerical graph.

About the student work: Three types of responses are seen in students' work: (1) those that correctly relate the change in speed to the amount of time traveled; (2) those that relate the straight line to a steady rate; and (3) those that see the negative and positive slope as going up and down a hill. An example of each type of response is included.

Student Work

Student Response A

In parts a and b the explanation relates the straight line to a steady "pace," not to an increase (a) or decrease (b) in speed over time. The relationship between speed and time is correct in part c.

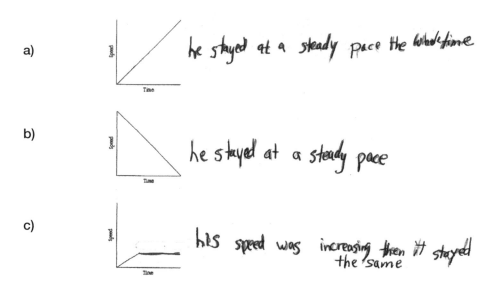

a)   he stayed at a steady pace the whole time

b)   he stayed at a steady pace

c)   his speed was increasing then it stayed the same

Student Response B

This explanation relates the direction of the line to the direction (up, down, straight), not to the rate.

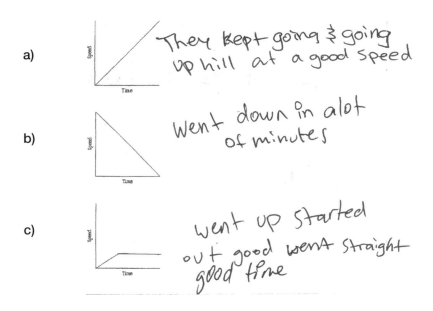

a) They kept going & going up hill at a good speed

b) Went down in alot of minutes

c) went up started out good went straight good time

Student Response C

This explanation relates the change of speed to increases in length of time elapsed, and appropriately interprets the horizontal line in part c as indicating that the speed stays the same over time.

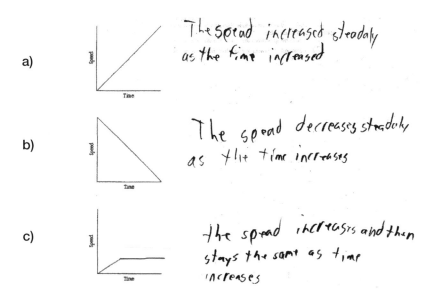

a) The speed increased steadaly as the time increased

b) The speed decreases steadaly as the time increases

c) the speed increases and then stays the same as time increases

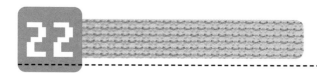

The liberty bicycle path is 36 miles long. The graph represents Anne's ride along the bike path one afternoon between 3:00 and 6:00 p.m.

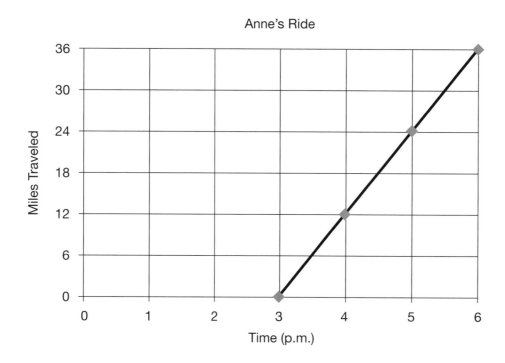

Anne's Ride

a. How can you tell from the graph that Anne was riding at the same speed for the whole trip?
b. At what speed was Anne riding?

Source: Balanced Assessment (Cambridge, Mass.: President and Fellows of Harvard College, 2000, p. 1 of 1, assessment task m016.doc)

About the mathematics: This item assesses students' ability to interpret the meaning of slope in the context of the situation.

Solution

a. The slope stays the same for whole trip (the distance traveled increases at same rate over time).
b. 12 miles per hour

Rubric (2 possible points): 1 point for each correct answer

Teacher note: This question is included to assess students' understanding of the meaning of the relationship between the distance traveled and time. This problem is included in contrast with question 21, which analyzes the relationship between time and speed.

About the student work: Students were generally successful on this item. Most responses referenced the straight line as evidence that Anne was riding at the same speed for the whole trip, and determined the speed at which she was riding.

Student Work

Student Response

The response references the straight line as evidence that Anne is riding at the same speed for the whole trip. The answer of 12 miles per hour is correct.

Part a

The line on the graph is straight.

Part b

12 mph

Geometry

THROUGH the study of geometry, students learn about geometric shapes and structures and how to analyze their characteristics and relationships. Through the use of geometric representations, students are often able to make sense of numeric and algebraic concepts. Geometric models are instrumental in helping students make sense of area and fractions, histograms and scatterplots foster insights about data, and coordinate graphs can serve to connect geometry and algebra (National Council of Teachers of Mathematics 2000, p. 41).

Middle school students should be engaged in building on previous geometric experiences. They should investigate relationships by drawing, measuring, visualizing, comparing, transforming, and classifying geometric objects. Their understanding of the properties of shapes should be such that, for example, they understand the defining characteristics of parallelograms and that a square is a special case of a rhombus and of a rectangle, each of which is a special cases of a parallelogram (NCTM 2000, p. 233).

Because proportional reasoning is a major concept for middle school students, they need to deepen and refine their understanding of congruence and similarity. Students in grades 6–8 should therefore have experiences that allow them to conjecture about the relationships between corresponding angles and corresponding sides of similar figures.

In grades 6–8 students are expected to investigate different transformations as they develop a strong understanding of line and rotational symmetry, scaling, and properties of polygons. In studying transformational geometry, students should encounter appropriate situations to help them understand and differentiate among lines of reflection, centers of rotation, and the position of an original figure and that of its image.

Visualizing and reasoning about spatial relationships are fundamental in geometry. These skills are best developed through experiences with dynamic software and either geometric models or isometric dot paper. Middle school students are expected to understand and conjecture about the relationships between two-dimensional representations and the corresponding three-dimensional objects. They should have experiences in interpreting or drawing different perspectives of buildings, and in drawing objects from geometric descriptions.

Meaningful experiences in geometry are important aspects of being successful in mathematics. Therefore, the use of good instructional practices that develop students' geometric concepts and language is important. Thoughtful assessments—of which the collection in this volume is meant to be a sampling rather than full coverage of the geometry topics in the middle-grades curriculumum—will help teachers adapt and improve their instructional practices as well as motivate students to learn.

Geometry Assessment Items

Standard: Analyze characteristics and properties of two- and three-dimensional geometric shapes and develop mathematical arguments about geometric relationships

Expectation: Precisely describe, classify, and understand relationships among types of two- and three- dimensional objects using their defining properties

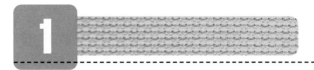

A quadrilateral MUST be a parallelogram if it has—

a. one pair of adjacent sides equal.
b. one pair of parallel sides.
c. a diagonal as an axis of symmetry.
d. two adjacent angles equal.
e. two pairs of parallel sides.

> **Source:** Adapted from TIMSS Population 2 Item Pool (The Hague: International Association for the Evaluation of Educational Achievement [IEA], 1994, p. 13)
> **About the mathematics:** This item reinforces the fact that a parallelogram is a quadrilateral with two pairs of parallel sides.
> **Solution:** The correct response is e.

Distractors

   a. This response defines a no specific quadrilateral.

   b. This response is one definition of a trapezoid.

   c. This response defines a rhombus or square.

   d. This response defines a kite, rhombus, or square.

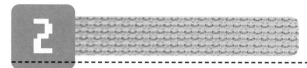

If no other sides or angles are congruent, which best describes the figure? How do you know?

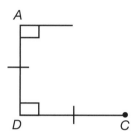

Source: Adapted from Massachusetts Comprehensive Assessment System (MCAS): Grade 10 (Malden, Mass.: Massachusetts Department of Education, 2004, p. 211). http://www.madoe.mass.edu/mcas.

About the mathematics: This item reinforces one common definition of a trapezoid: a quadrilateral with one pair of opposite sides parallel.

Solution: The completed figure would be a trapezoid. The two right angles require the opposite sides to be parallel, so no other sides are parallel. Only one pair of adjacent sides is congruent.

Rubric (1 possible point)

   1 point: Gives correct response and a complete explanation

   0.5 points: Gives correct response but an incomplete explanation or no explanation

   0 points: Gives incorrect answer or makes no attempt at an answer

Teacher note: An incomplete explanation would include some but not all the necessary conclusions.

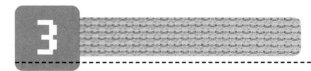

Lines m and n do not intersect. Line p intersects line m to form a right angle. Which of the following is NOT true?

a. Lines p and n are parallel.
b. Lines m and n are parallel.
c. Lines p and m are perpendicular.
d. All three statements are true.

> **Source:** Nova Scotia Department of Education Assessment Program
> **About the mathematics:** To be successful with this item, students need to understand that if parallel lines are cut by a transversal, the corresponding angles and alternate interior angles are equal.
> **Solution:** a
> **Distractors**
> a. We are given that m and n do not intersect.
> b. We are given that p and m form a right angle.
> c. We are given that m and n are parallel and that p intersects m; therefore, p must intersect n.

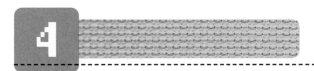

Luis claimed that a figure must be a square if it is a parallelogram and has all its sides the same length. Is Luis correct? Defend your decision with a written argument.

> **Source:** Adapted from Nova Scotia Department of Education Assessment Program
> **About the mathematics:** This item assesses students' understanding that a square is a parallelogram with equal sides and containing one right angle.
> **Solution:** Luis is incorrect. Luis' explanation does not include the requirement that a square must contain at least one right angle.
> **Teacher note:** Along with stating that to be a square, a parallelogram must have all sides equal, one does not need to state that it has four right angles; however, that statement is not incorrect.

5

Five geometric terms are listed in alphabetical order below.

Equilateral Triangle   Rhombus   Right Isosceles Triangle   Square   Trapezoid

Four different shapes are represented below.

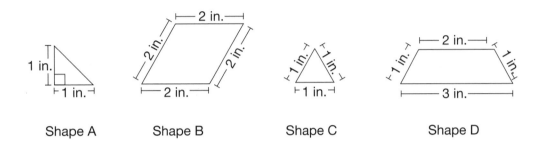

Shape A   Shape B   Shape C   Shape D

a. Which one of the geometric terms listed above most accurately describes shape A? Explain how you know that your answer is correct.
b. Which one of the geometric terms listed above most accurately describes shape B? Explain how you know that your answer is correct.
c. Which one of the geometric terms listed above most accurately describes shape C? Explain how you know that your answer is correct.
d. Which one of the geometric terms listed above most accurately describes shape D? Explain how you know that your answer is correct.

Rubric (4 possible points)

4 points: Correctly identifies all four shapes and includes accurate properties for each shape

3 points: Correctly identifies three shapes and includes accurate properties for each shape, or correctly identifies four shapes but misses properties of two shapes, or correctly identifies two shapes but includes accurate properties for four shapes

2 points: Correctly identifies two shapes and includes accurate properties for two shapes, or correctly identifies all four shapes but neglects to include or erroneously includes properties for four shapes

1 point: Correctly identifies one shape and includes accurate properties for one shape, or correctly identifies two shapes

0 points: Incorrectly identifies each shape and lists inaccurate properties for each shape, or makes no attempt at an answer

Source: Adapted from Massachusetts Comprehensive Assessment System (MCAS): Grade 6 (Malden, Mass.: Massachusetts Department of Education, 2004, p. 40, item 10)

About the mathematics: Students must associate geometric figures with their appropriate names from among those listed, on the basis of their knowledge of the definitions of the respective figures.

Solution

    a. Shape A is a right isosceles triangle.

    b. Shape B is a rhombus.

    c. Shape C is an equilateral triangle.

    d. Shape D is a trapezoid.

Standard: Analyze characteristics and properties of two- and three-dimensional geometric shapes and develop mathematical arguments about geometric relationships

Expectation: Understand relationships among the angles, side lengths, perimeters, areas, and volumes of similar objects

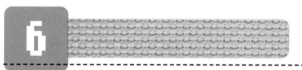

This sketch shows two similar triangles, *ABC* and *PQR*. Triangle *ABC* has an area of 20 square units, and its altitude, $\overline{CD}$, is equal to 4 units. Triangle *PQR* is similar to triangle *ABC*, and its altitude, $\overline{RS}$, is equal to 8 units. What is the area of triangle *PQR*?

    a. 40 square units

    b. 80 square units

    c. 160 square units

    d. 640 square units

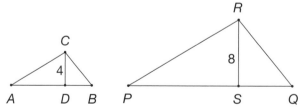

About the mathematics: To be successful with this item, students need to realize that the similarity factor is 2 but that it must be applied to two lengths (the base and the height), so the area increases by a factor of 4.

Solution: The correct answer is b.
Distractors
a. This response indicates that the student did not remember that the ratio of similitude for area is a squared quantity.
c. In this response, the student has multiplied 8 times the area.
d. In this response, the student has multiplied all the numbers: $4 \times 8 \times 20$.

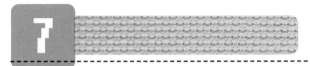

Can a parallelogram have two 45° and two 75° angles? Why or why not?

> **Source:** *Connected Mathematics Project,* "Shapes and Designs" (Upper Saddle River, N.J.: Prentice Hall, 1996, p. 137)
> **About the mathematics:** This item assesses students' understanding that the sum of the interior angles in a quadrilateral is 360 degrees.
> **Solution:** No, because the sum of the angle measures is only 240 degrees. The sum of the angle measures of a parallelogram must be 360 degrees.

How many triangles of the shape and size of the triangle A can the trapezoid be divided into?

a. three
b. four
c. five
d. six

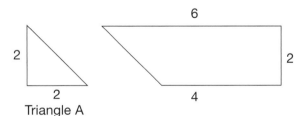

> **Source:** TIMSS population 2 Item Pool (The Hague: International Association for the Evaluation of Educational Achievement [IEA], 1994, p. 103)

About the mathematics: This item involves tessellating a figure with a given figure when dimensions of each are given.

Solution: The correct response is c.

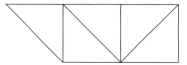

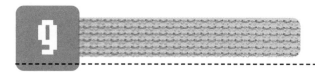

Figure 1 is a special kind of pentagon in which *ABCD* is a square and *E* is the midpoint of the diagonal *BD;* then square *DECF* is drawn to make the pentagon *ABCFD*. Figure 2 continues drawing similar figures, starting with square *ABCD*, then adding square *ECFD*, and so on.

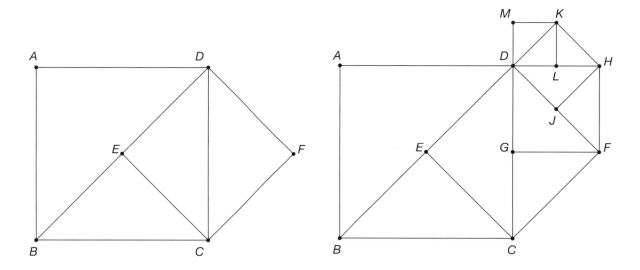

a. Find three pentagons similar to *ABCFD.*
b. If $\overline{DM}$ is 2 cm long, how long is $\overline{DA}$?

About the mathematics: This item requires students to repeat a described procedure and recognize the properties of the similar figures formed at each step.

Solution

a. Pentagons *ECFHD, GFHKD,* and *JHKMD*

b. $DM = 2$, $DM = DL$, $DL = \frac{1}{2}DH$; therefore $DH = 4$, $DH = DG$, $DG = \frac{1}{2}DC$, $4 = \frac{1}{2}DC$. So $DC = 8$, and $DC = AD$; therefore $AD = 8$.

Or:

If $DM = 2$, then $DK = 2\sqrt{2}$. If $DK = 2\sqrt{2}$, then $DH = 4$. If $DH = 4$, then $AD = 8$.

Rubric (4 possible points)

4 points: Correctly names three pentagons and correctly finds the value of *DA*

3 points: Correctly names three pentagons but incorrectly computes the value of *DA*, or correctly names two pentagons and correctly finds the value of *DA*

2 points: Correctly names two pentagons but incorrectly computes the value of *DA*, or correctly names one pentagon and correctly finds the value of *DA*

1 point: Correctly names one pentagon but incorrectly computes the value of *DA*, or correctly names no pentagons but correctly finds the value of *DA*

0 points: Makes no attempt at a solution or gives all incorrect answers

About the student work: In general the student responses on the pilot test were correct. The most frequent errors were exemplified by student responses A and C and by no attempt being made.

Student Work

Student Response A

In this response the student identifies three pentagons but makes some errors in ordering the vertices in part a; part b is correct, although no work is shown.

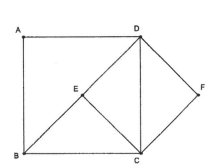

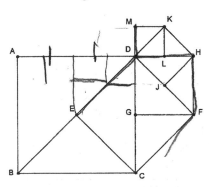

a.   Find three pentagons similar to ABCFD. dg fhk/ecfdh/DJhkm

b.   If segment DM is 2 cm long, how long is segment DA ? 8cm long

Student Response B

This response exemplifies a successful solution.

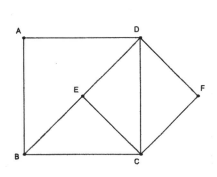

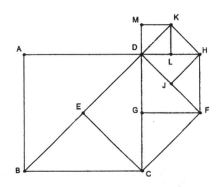

a.   Find three pentagons similar to ABCFD.
      ECFHD   GFHKD   JHKMD

b.   If segment DM is 2 cm long, how long is segment DA ?
      8 cm long

Student Response C

The work shows an incomplete part a; however, part b is correct.

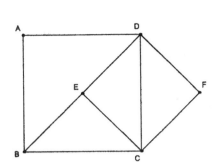
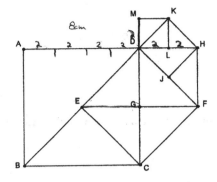

a.   Find three pentagons similar to ABCFD.
DGFHK,

b.   If segment DM is 2 cm long, how long is segment DA ?
8cm long

Standard: Analyze characteristics and properties of two- and three-dimensional geometric shapes and develop mathematical arguments about geometric relationships

Expectations: Create and critique inductive and deductive arguments concerning geometric ideas and relationships, such as congruence, similarity, and the Pythagorean relationship

---

10

This sketch shows a 6 × 10 rectangle with its diagonal drawn and another rectangle in the upper-left corner. Some lengths are shown in the sketch.

Find the length of x.

Explain how you know the length of x using mathematical language.

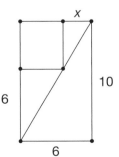

Solution: Since opposite sides of a rectangle are parallel, the diagonal is a transversal. On the basis of the properties of parallel lines cut by a transversal, we can determine that alternate interior angles are congruent and right angles are congruent, so the figure contains three similar triangles. From two of the triangles, we can show that the ratio of the sides is $\frac{4}{x} = \frac{10}{6}$. Therefore, $x = 2.4$.

Rubric (2 possible points)

2 points: Correctly finds the value of x and provides a mathematically correct explanation that includes the properties of similarity

1.5 points: Shows that the side lengths are proportional and provides a mathematically correct explanation that includes the properties of similarity but makes an arithmetic error in calculating the value of x

1 point: Correctly finds the value of x but neglects to provide an explanation, or offers an explanation that does not reference similar triangles, or attains an incorrect value for x because of an arithmetic error but provides a correct explanation

0.5 points: Correctly finds the value of x but provides an incomplete explanation

0 points: Gives an incorrect answer with either an incomplete explanation or no explanation

About the student work: Students generally had difficulty in interpreting this problem, as exemplified by the three examples of student work.

Student Work

Student Response A

This response exemplifies an error of assuming that the small rectangle is a square.

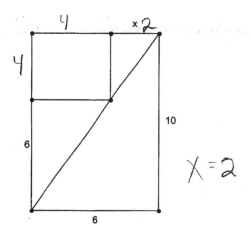

Student Response B

This response identifies the rectangle but does not include a proportion for the triangles.

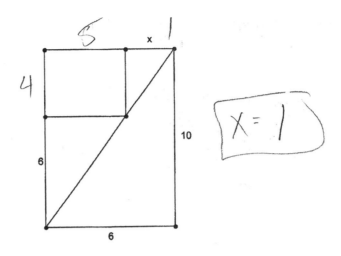

Student Response C

In this response, the student shows an attempt to use the Pythagorean theorem but misses the identification of the ratio of the triangles.

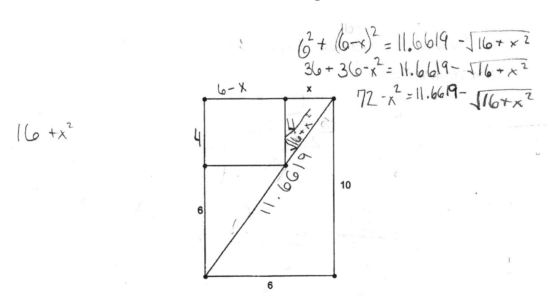

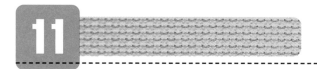

The area of the largest semicircle is 30π.

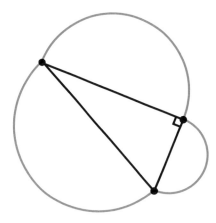

Find the sum of the areas of the smaller two circles. Explain how you know.

> **About the mathematics:** This item requires students to apply the Pythagorean theorem.
>
> **Solution:** 30π. Because the diameters of the circles are the legs of the triangles, and because the Pythagorean theorem states that $a^2 + b^2 = c^2$, the sum of the areas of the two smaller circles equals the area of the larger circle.

Standard: Specify locations and describe spatial relationships using coordinate geometry and other representational systems

Expectation: Use coordinate geometry to represent and examine the properties of geometric shapes

12

Which of the following is an equation for a line that is *not* parallel to the line that has the following equation?

$$-x + 3y - 6 = 0$$

a. $y = \frac{1}{3}x - 2$
b. $y = \frac{1}{3}x + 2$
c. $-x + y - 2 = 0$
d. $x - 3y + 6 = 0$

> **About the mathematics:** This item requires students to know that the slopes of parallel lines must be equal and to determine the slope given an equation.
> **Solution:** The correct response is c.
> **Distractors**
> > a. The line represented by this equation has the same slope as the given line ($\frac{1}{3}$).
> > b. The line represented by this equation is an equivalent form of the given line.
> > d. The line represented by this equation has the same slope as the given line ($\frac{1}{3}$).

13

- Point $A\,(0, 3)$ is on the y-axis.
- Point $B\,(4, 0)$ is on the x-axis.
- Points A and B are 5 units apart.
- Lines parallel to $\overline{AB}$ make triangles that are similar to triangle AOB.

Find the points on the y-axis and x-axis that are 30 units apart, and make a line parallel to $\overline{AB}$. Explain how you know. Be sure to use mathematical language in your explanation.

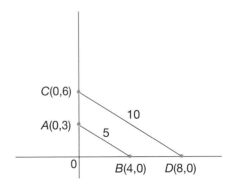

About the mathematics: Students need to be able to sketch the triangle with points 30 units apart and find the coordinates of the points on the axes.
Solution: The points are (0, 18) and (24, 0). Because the triangle formed would be similar to triangle *AOB*, we know that $3/5 = y/30$ and $4/5 = x/30$; hence the ordered pairs of the endpoints are (0, 18) and (24, 0).
About the student work: In general, students' responses to this item on the pilot test showed an understanding of the problem. The most frequent errors occurred in computing the coordinate values.

Student Work

Student Response A

In this response the student shows an understanding of the question but identifies the wrong pattern for the coordinates on the axes.

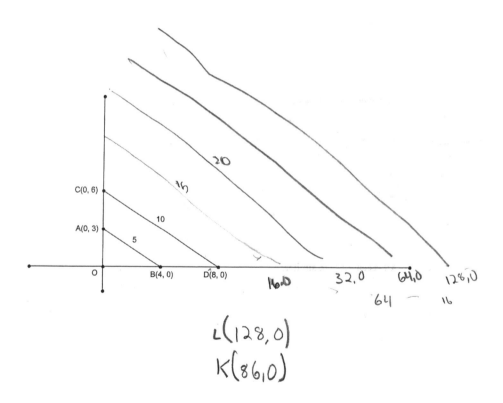

Student Response B

In this response the student indicates a pattern but draws only the final line. An error is made in naming the points on the axes.

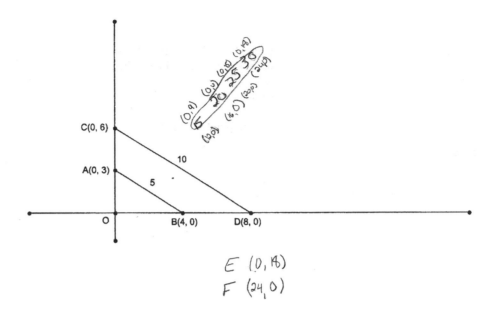

$$E\ (0, 18)$$
$$F\ (24, 0)$$

Student Response C

This response is an example of finding the pattern without drawing any of the lines, yet answering the question correctly; however, the notation for the final ordered pair is written incorrectly

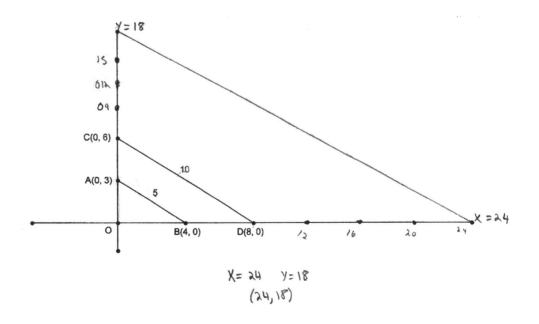

$$X = 24 \quad Y = 18$$
$$(24, 18)$$

Standard: Apply transformations and use symmetry to analyze mathematical situations

Expectation: Describe sizes, positions, and orientations of shapes under informal transformations such as flips, turns, slides, and scaling

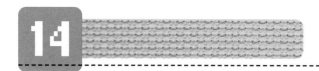

In parts a–c, explain in your own words what would happen to any figure if you transformed it using the given rule.

a. $(3x, 6y)$
b. $(x + 2, y + 1)$
c. $(2x, 2y + 5)$

Source: *Connected Mathematics Project,* "Stretching and Shrinking" (Upper Saddle River, N.J.: Prentice Hall, 1996, p. 99)

About the mathematics: This item assesses students' ability to identify the effects of transforming a figure according to a given rule.

Solution

a. The horizontal lengths would increase by a factor of 3, and the vertical lengths would increase by a factor of 6.

b. The figure would stay the same size, but it would move to the right 2 units and up 1 unit.

c. The figure would increase by a scale factor of 2 and would move up 5 units.

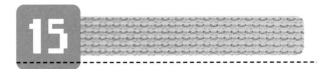

Which of the following represents the coordinates of the vertices after a rotation of 180° about the origin?

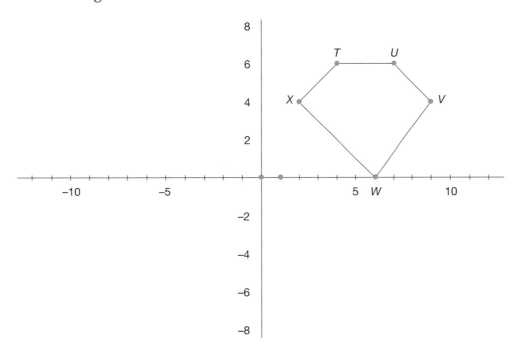

a. $T'(-4, -6)$, $U'(-7, -6)$, $V'(-9, -4)$, $W'(-6, 0)$, $X'(-2, -4)$
b. $T'(-4, 6)$, $U'(-7, 6)$, $V'(-9, 4)$, $W'(-6, 0)$, $X'(-2, 4)$
c. $T'(4, -6)$, $U'(7, -6)$, $V'(9, -4)$, $W'(6, 0)$, $X'(2, -4)$
d. $T'(6, 4)$, $U'(6, 7)$, $V'(4, 9)$, $W'(0, 6)$, $X'(4, 2)$

Source: Adapted from Nova Scotia Department of Education Assessment Program

About the mathematics: This item concerns rotating a figure about a point and locating points on a Cartesian plane.

Solution: The correct response is a.

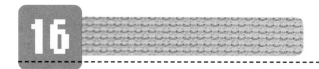

Geometry is a must for computer graphics. Moving a figure smoothly around the computer screen can be done by performing many small slides (translations) very quickly, one after another. A translation that moves every point in the plane 2 units to the right and 3 units down can be represented by $T(+2, -3)$. The cartoon figure shown below has the tip of its nose at the point (80, 100).

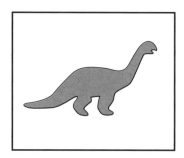

Describe a single translation that would transform the cartoon figure so that the resulting image would have the tip of its nose at the point (95, 70).

Source: Adapted from Nova Scotia Department of Education Assessment Program

About the mathematics: This item concerns translating a point and identifying its new position on the Cartesian plane.

Solution: The correct translation is 15 units right and 30 units down, $T(+15, -30)$.

Which of the following two shapes can be combined to form a tessellation?

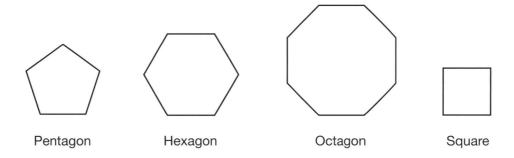

Pentagon   Hexagon   Octagon   Square

a. The pentagon and the hexagon
b. The octagon and the square
c. The pentagon and the octagon
d. The hexagon and the square

Source: Adapted from Nova Scotia Department of Education Assessment Program

About the mathematics: This item requires an understanding of tessellation of a plane.

Solution: The correct response is b.

Standard: Apply transformations and use symmetry to analyze mathematical situations

Expectation: Examine the congruence, similarity, and line or rotational symmetry of objects using transformations

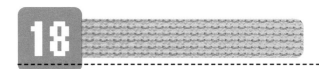

The figure below is one-half of a symmetric figure with its line of symmetry.

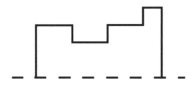

Which is the other half of the figure?

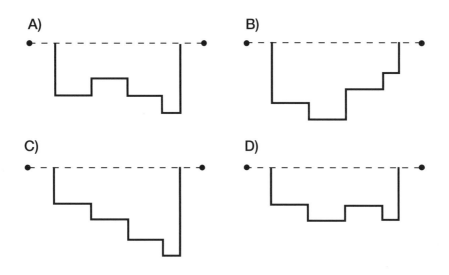

Source: Nova Scotia Department of Education Assessment Program
About the mathematics: This item involves applying line symmetry and understanding the concept of a symmetric figure.
Solution: The correct response is A.

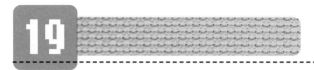

$\triangle ABC$ and $\triangle DEF$ are shown on the grid below.

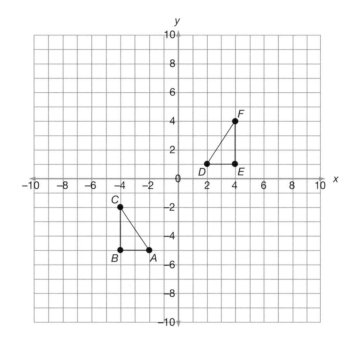

Which of the following transformations will map $\triangle ABC$ onto $\triangle DEF$?

a. Reflect $\triangle ABC$ over the y-axis, and shift up 6 spaces.
b. Reflect $\triangle ABC$ over the x-axis, and shift up 6 spaces.
c. Reflect $\triangle ABC$ over the y-axis, and shift down 6 spaces.
d. Reflect $\triangle ABC$ over the y-axis, reflect over the x-axis, and shift down 4 spaces.

Source: Adapted from Massachusetts Comprehensive Assessment System (MCAS): Grade 8 (Malden, Mass.: Massachusetts Department of Education, 2002, p. 236, item 18)

About the mathematics: This item combines translating a figure and reflecting it over a line in a specific order.

Solution: The correct response is a.

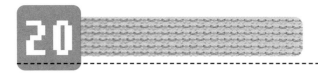

Describe two different sets of transformations that would move square *PQRS* onto square *WXYZ.*

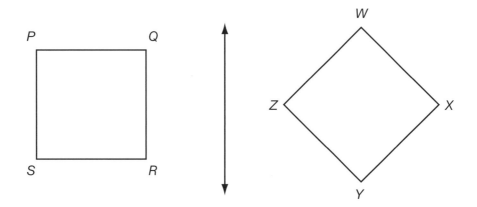

Source: *Connected Mathematics Project*, "Kaleidescopes, Hubcaps, and Mirrors" (Upper Saddle River, N.J.: Prentice Hall, 1996, p. 191)

About the mathematics: This item involves using transformation to show congruence.

Solution: Answers may vary, but the following is one possible solution: Reflect square *PQRS* over the line, and then rotate it 45° about its center point. Rotate square *PQRS* 135°, and then translate it until vertex *P* matches vertex *Y.*

Standard: Use visualization, spatial reasoning, and geometric modeling to solve problems

Expectation: Draw geometric objects with specified properties, such as side lengths or angle measures

21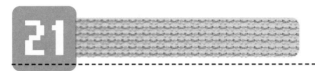

The following list contains the results Yvonne got when she rolled three number cubes simultaneously and recorded the numbers appearing on the three faces.

a. (1, 3, 3)
b. (5, 3, 4)
c. (4, 6, 2)
d. (4, 3, 2)

If each set of three numbers was used as the lengths of the sides, could Yvonne always make a triangle? Explain how you know. Be sure to use mathematical language in your explanation.

> **About the mathematics:** This item involves an application of the triangular inequality theorem.
>
> **Solution:** The student should indicate that to form a triangle, the sum of any two of the numbers must be greater than the third. By using the triangular inequality, the student should conclude that the set (4, 6, 2) will not form a triangle.

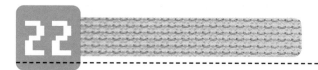

The perimeter of a triangle is 15. Make a list of all possible whole-number side lengths for the triangle.

About the mathematics: This item involves an application of the triangular inequality theorem.

Solution: Seven triangles are possible: 5-5-5, 5-6-4, 5-3-7, 6-6-3, 4-4-7, 7-7-1, and 7-6-2.

Standard: Use visualization, spatial reasoning, and geometric modeling to solve problems

Expectation: Use two-dimensional representations of three-dimensional objects to visualize and solve problems such as those involving surface area and volume

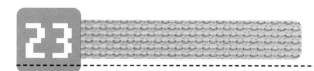

A *polycube* is a shape made by attaching shapes to a cube so that each additional cube shares at least one entire face with another. The three diagrams shown below represent the shadows cast by a polycube when it is illuminated from the front, top, and side by sunlight.

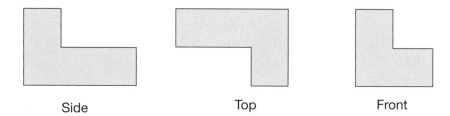

Side   Top   Front

Which of the following correctly represent the polycube that cast the shadows indicated above?

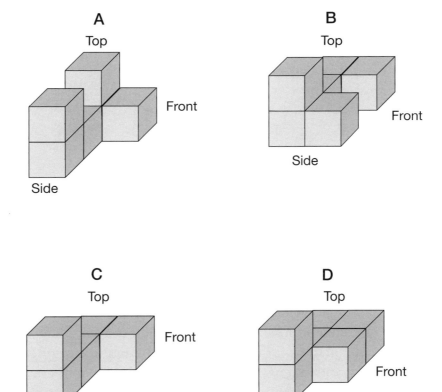

A
Top
Front
Side

B
Top
Front
Side

C
Top
Front
Side

D
Top
Front
Side

Source: Adapted from Nova Scotia Department of Education Assessment Program

About the mathematics: This item involves visualizing a three-dimensional object given three views in two-dimensions

Solution: The correct response is C.

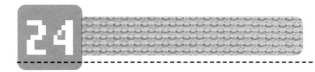

The pattern shown below is for a rectangular prism. The lengths of the line segments in the pattern were chosen so that the pattern could be folded along the dotted lines into the prism shown.

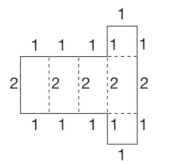

a. Make a sketch of a pattern for a triangular prism. Label **each** line segment with a length that will make it possible to fold the pattern into the triangular prism.

b. Make a sketch of a pattern for a cylinder. Label **each** line segment and diameter in your pattern with a length that will make it possible to create the cylinder from the pattern.

> **Source:** Massachusetts Comprehensive Assessment System (MCAS): Grade 8 (Malden, Mass.: Massachusetts Department of Education, 2001, p. 295, item 8)
> **About the mathematics:** This item involves visualizing a pattern for folding a figure given the name of a figure, as well as accurately identifying the dimensions of the figure.
> **Solution:** Answers will vary in terms of the dimensions.

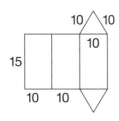

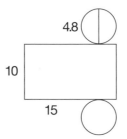

Rubric (4 possible points)

4 points: Shows comprehensive sense of spatial relationships by making accurate sketches of patterns for two three-dimensional geometric figures and labeling the lengths of the edges appropriately

3 points: Shows good sense of spatial relationships by making accurate sketches of patterns for two three-dimensional geometric figures and labeling the lengths of edges; may omitt one or two significant measures

2 points: Shows partial sense of spatial relationships by inconsistently making sketches of patterns for two three-dimensional geometric figures or inconsistently labeling the lengths of the edges

1 point: Shows limited sense of spatial relationships by making major errors in sketches and labeling

0 points: Give an incorrect response or a response containing some correct work that is irrelevant to the skill or concept being measured

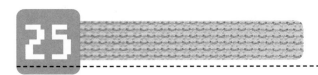

A 3-inch cube has its top and two adjacent vertical faces painted blue and its bottom and the other two adjacent vertical faces painted red. After the paint dried, the cube was cut into 27 one-inch cubes

a. How many of the 27 one-inch cubes would have three faces painted blue and three faces without paint?

b. How many of the 27 one-inch cubes would have two faces painted red and four faces without paint?

c. How many of the 27 one-inch cubes would have one face painted red, one face painted blue, and four faces without painted?

d. How many of the 27 one-inch cubes would have two faces painted blue, one face painted red, and three faces without paint?

e. How many of the 27 one-inch cubes would have one face painted red and five faces without paint?

f. How many of the 27 one-inch cubes would have 6 faces unpainted?

Source: Adapted from "Activities: Discovery Using Cubes" by Robert Reys (*Mathematics Teacher* 67 [January 1974]: 47–50)

About the mathematics: This item involves visualization of the dissection of a cube.

Rubric (6 possible points): 1 point for each correct response
Solution

    a. One small cube will have three blue faces and three unpainted faces.

    b. Three small cubes will have two red faces and four unpainted faces.

    c. Six small cubes will have one red face, one blue face, and four unpainted faces.

    d. Three small cubes will have two blue faces, one red face, and three unpainted faces.

    e. Three small cubes will have one red face and five unpainted faces.

    f. One small cube will have six unpainted faces.

Standard: Use visualization, spatial reasoning, and geometric modeling to solve problems

Expectation: Recognize and apply geometric ideas and relationships in areas outside the mathematics classroom, such as art, science, and everyday life

26

A person standing 10 feet from a 20-foot streetlight has a shadow 4 feet long.

How tall is the person?

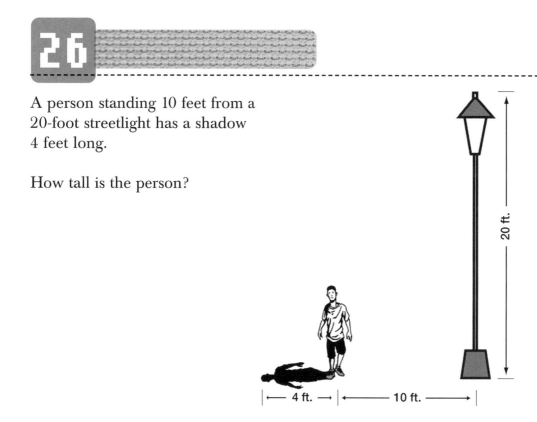

20 ft.

|← 4 ft. →|←——— 10 ft. ———→|

About the mathematics: This item involves the application of similar triangles.

Solution: A correct solution includes a proportion correctly indicating the following:

$$\frac{\text{height of streetlight}}{\text{height of person}} = \frac{(\text{length of shadow}) + (\text{distance from streetlight})}{\text{length of shadow of person}},$$

that is,

$$\frac{20}{x} = \frac{14}{4},$$

$$x = 5.7 \text{ feet},$$

or approximately 5 feet, 8 inches.

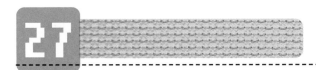

Describe a strategy involving shadows and similar triangles that you could use to find the height of your school building. Be sure to tell what measurements and calculations you would make.

Source: Nova Scotia Department of Education Assessment Program

About the mathematics: This item involves making a precise plan for solving a problem without actually solving it.

Solution: Solutions will vary. One suggestion might be to stand in line with the building so that the sun casts your shadow in the same direction as that of the building. You could then measure the distance to the end of each shadow. Knowing or measuring your height would then allow you to determine the height of the building by using similar triangles.

Measurement

STUDENTS in the middle grades have had many diverse experiences with measurement both from prior classwork and from using measurement in their daily lives. At this grade level, students are expected to build on and formalize their measurement experiences. In these grades appropriate tasks for students include working with measurable attributes, such as length, area, and volume; correctly applying appropriate units of measure; and working flexibly within systems of measurement. Middle school students should have many opportunities for choosing and using compatible units for the attributes being measured, estimating measurements, selecting appropriate units and scales on the basis of the precision desired, and solving problems involving the perimeter and area of two-dimensional shapes and the surface area and volume of three-dimensional objects.

Students at this level are also expected to become proficient at measuring angles and using ratio and proportion to solve problems involving scaling, similarity, and derived measures. Separating measurement from geometry is often difficult to do, because many of the measurement topics studied are closely linked with the concepts that students study in geometry. Measurement is also tied to ideas and skills in number, algebra, and data analysis in such topics as distance-velocity-time relationships and data collected by direct or indirect measure (NCTM 2000).

The problems in this chapter span the Measurement Standards and Expectations outlined in *Principles and Standards for School Mathematics* (NCTM 2000). Many of the items address more than one standard or more than one expectation. For those items, multiple standards and expectations are listed. The set of problems included here is by no means complete but, rather, was selected to show how certain concepts can be assessed.

Measurement Assessment Items

Standard: Understand measurable attributes of objects and the units, systems, and processes of measurement

Expectations: Understand both metric and customary systems of measurement; understand relationships among units and convert from one unit to another within the same system

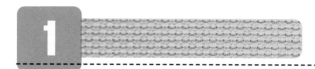

In the 4 by 220 relay race, each sprinter runs 220 yards. How far does each sprinter run in miles? Show all your work.

> **About the mathematics:** This item involves conversion from one unit to another within a system of measure.
>
> **Solution:** $1/8$ of a mile. Since 1 mile is 1760 yards, calculate $^{220}/_{1760}$, giving $1/8$ of a mile.

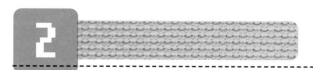

Name the customary and metric units that you could use to measure the following items. For example, for a DVD player you could measure the length, width, or height in inches or centimeters; you could measure the weight or mass in pounds or kilograms; and you could measure the surface area in square inches or square centimeters.

	Customary Unit	Metric Unit
a. Can of soda		
b. A car		
c. A trip from Boston to San Diego		
d. The water used in your school each day		
e. The area of your state		

About the mathematics: This item requires appropriate application of customary and metric units to describe everyday objects or experiences.

Solution

    a. Ounces or milliliters

    b. Length, width, or height in feet or meters; weight or mass in pounds, tons, or kilograms

    c. Miles or kilometers, or in terms of time, as days for a car trip or hours for an airplane trip

    d. Volume in gallons or liters

    e. Area in square miles or square kilometers

Standards: Understand measurable attributes of objects and the units, systems, and processes of measurement; apply appropriate techniques, tools, and formulas to determine measurements

Expectations: Understand both metric and customary systems of measurement; understand relationships among units and convert from one unit to another within the same system; select and apply techniques and tools to accurately find length, area, volume, and angle measures to appropriate levels of precision; solve simple problems involving rates and derived measurements for such attributes as velocity and density

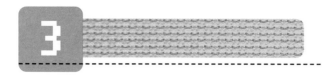

A news report about lottery winnings stated, "Saturday's $43 million Lotto Jackpot equals a trail of $1 bills that would stretch 4100 miles, from New York City to San Francisco and back to Glacier National Park in Montana."

a. A $1 bill is about 6 inches long. How many $1 bills are needed to make a trail 1 mile long?
b. How many $1 bills are needed to make a trail 4100 miles long?

Source: *Connected Mathematics: "Data around Us"* (Upper Saddle River, N.J.: Prentice Hall, 1996, p. 33, item 11)

About the mathematics: This item involves converting inches to miles in standard measure.

Solution

    a. 1 mile = 5280 feet. Each foot will have 2 bills, so 5280 × 2 = 10,560 bills.

    b. 10,560 bills/mile × 4100 miles = 43,296,000 bills.

Standards: Understand measurable attributes of objects and the units, systems, and processes of measurement; apply appropriate techniques, tools, and formulas to determine measurements

Expectations: Understand both metric and customary systems of measurement; understand relationships among units and convert from one unit to another within the same system; use common benchmarks to select appropriate methods for estimating measurements; solve simple problems involving rates and derived measurements for such attributes as velocity and density

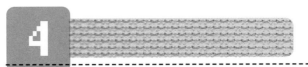

In 1999 Matt Anderson, a pitcher with major league baseball's Detroit Tigers, threw the fastest pitch ever recorded up to that date. A radar gun clocked the pitch at an average of 103 miles per hour. If the distance from the pitching rubber to home plate is 60 feet 6 inches, approximately how long does the ball take to get to the batter after it leaves Anderson's hand?

Source: Adapted from A Fast Trip, http://mathforum.org/midpow/ solutions/19980511.midpow.html (accessed March 14, 2005)

About the mathematics: This item involves conversion from one unit to another within a system of measure and application of the formula $d = r \times t$.

Solution: Approximately 0.0066 minute, or 0.4 second. Convert 60 feet 6 inches to 60.5 feet. A mile is 5280 feet, so the ball must travel

$$60.5 \text{ ft.} \div 5280 \text{ ft./mi.} \approx 0.0114 \text{ mi.}$$

Then calculate

$$0.0114 \text{ mi.} \div 103 \text{ mi./hr.} \approx 0.000111 \text{ hr.,}$$
$$0.000111 \times 60 \approx .0066 \text{ min.,}$$
$$0.0066 \times 60 \approx 0.4 \text{ sec.}$$

Standards: Understand measurable attributes of objects and the units, systems, and processes of measurement; apply appropriate techniques, tools, and formulas to determine measurements

Expectations: Understand both metric and customary systems of measurement; understand relationships among units and convert from one unit to another within the same system; understand, select, and use units of appropriate size and type to measure angles, perimeter, area, surface area, and volume

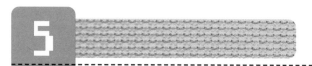

The muscles in the human eye move an estimated 100,000 times or more each day. (Source: *Guiness Book of Records 1993,* edited by Peter Matthews [New York: Bantam Books, 1993])

a. About how many times do your eye muscles move in one year?
b. About how many movements do your eye muscles make in one second?
c. In about how many days will your eye muscles make 1,000,000 movements?

Source: Adapted from *Connected Mathematics Project,* "Data around Us" (Upper Saddle River, N.J.: Prentice Hall, 1996)

About the mathematics: This item involves an application of rates within one system of measure.

Solution
   a. $100,000 \times 365 = 36,500,000$ times.
   b. $100,000 \div 24 \div 60 \div 60 \approx 1.16$ movements per second.
   c. About ten days

Standards: Understand measurable attributes of objects and the units, systems, and processes of measurement; apply appropriate techniques, tools, and formulas to determine measurements

Expectations: Understand both metric and customary systems of measurement; understand, select, and use units of appropriate size and type to measure angles, perimeter, area, surface area, and volume; select and apply techniques and tools to accurately find length, area, volume, and angle measures to appropriate levels of precision; develop and use formulas to determine the circumference of circles and the area of triangles, parallelograms, trapezoids, and circles and develop strategies to find the area of more-complex shapes; develop strategies to determine the surface area and volume of selected prisms, pyramids, and cylinders

Suppose you want to put a wooden border around the top and sides of your classroom door. If your classroom has a single door, not a double door, explain how you could estimate the amount of wood you will need. Show all your work, and identify any assumptions you made.

Source: Adapted from Nova Scotia Department of Education Assessment Program

About the mathematics: This item involves the ability to gauge how tall a standard door is and to recognize the need for the length of two sides and one top.

Solution: The door is approximately 7 feet high and 4 feet wide, so about 18 feet of wood is needed; but because some waste might occur when cutting the wood, a safer estimate might be 20 to 22 feet.

Standards: Understand measurable attributes of objects and the units, systems, and processes of measurement; apply appropriate techniques, tools, and formulas to determine measurements

Expectations: Understand both metric and customary systems of measurement; understand, select, and use units of appropriate size and type to measure angles, perimeter, area, surface area, and volume; select and apply techniques and tools to accurately find length, area, volume, and angle measures to appropriate levels of precision; develop and use formulas to determine the circumference of circles and the area of triangles, parallelograms, trapezoids, and circles and develop strategies to find the area of more-complex shapes; solve problems involving scale factors, using ratio and proportion

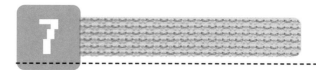

Lawrence Hargrave is noted for his designs for box kites. In his box kites, squares and rectangles form open rectangular prisms.

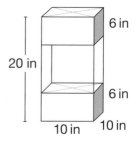

a. What is the surface area of the prism?
b. What fraction of the 10-inch-by-10-inch-by-20-inch prism is covered?

Source: Adapted from *Middle Grades Maththematics: Vol. 2* (Evanston, Ill.: McDougall Littell, 1999)

About the mathematics: This problem requires the use appropriate units of measure to find surface area.

Solution

a. The surface area is 2(10 in. × 10 in.) + 4(20 in. × 10 in.), which results in a total of 1000 square inches.
b. The covered surface area is 4(10 in. × 6 in.) + 4(10 in. × 6 in.), for a total of 480 square inches. The fraction that is covered is $^{480}/_{1000}$, or $^{48}/_{100}$, or $^{12}/_{25}$ square inches.

Standards: Understand measurable attributes of objects and the units, systems, and processes of measurement; apply appropriate techniques, tools, and formulas to determine measurements

Expectations: Understand both metric and customary systems of measurement; understand relationships among units and convert from one unit to another within the same system; understand, select, and use units of appropriate size and type to measure angles, perimeter, area, surface area, and volume; use common benchmarks to select appropriate methods for estimating measurements; develop and use formulas to determine the circumference of circles and the area of triangles, parallelograms, trapezoids, and circles and develop strategies to find the area of more-complex shapes; solve problems involving scale factors, using ratio and proportion

8

Mo has just bought a diner and is planning to remodel it before opening. The plan of the diner is drawn to a scale of 1 inch to 4 feet. Mo has asked your advice on how best to locate new tables and chairs, how many to buy, and how much the furniture will cost. She plans to order a supply of rectangular tables, circular tables,

Chair	**Round Table**	**Rectangular Table**
Seat base: 16 square inches	Diameter: 48 inches	Top: 60 inches × 30 inches
$199.99 each	$449.99 each	$569.99 each

and chairs. She wants the diner to hold as many people as possible, with adequate space for movement of customers and staff. She wants the diner to feel attractive and welcoming. Your task is to devise a plan for table layouts for her.

You will need to—
- investigate how tables and chairs will fit, using the floor plan of the diner, then
- draw one layout to scale on plans of the diner;
- say how much the furniture will cost for the recommended plan.

Your layouts should show the exact location of each table and how the chairs should be positioned. You may find grid paper and scissors helpful during your investigations.

> **Source:** Adapted from a task designed and developed by MARS: Mathematics Assessment Resource Service, http://www.educ.msu.edu/MARS/ (accessed March 14, 2005)
>
> **About the mathematics:** For this item, students interpret scale plans, draw a plan to scale, convert units, make recommendations, and determine costs.
>
> **Solution:** Students' responses will vary.
>
> **Rubric (3 possible points)**
>
> *3 points:*   Level 3—meets all expectations
>
> *2 points:*   Level 2—meets some expectations
>
> *1 point:*   Level 1—falls below expectations
>
> **Teacher note:** In this planning-and-design task, the student has to produce and comment on an alternative workable layout for tables and chairs in a diner to maximize the number of customers that can be accommodated comfortably. Teachers may want students to work individually or in pairs to start with, then devise individual plans to finish. The materials needed are any inch-based grid paper, ruler, pencil, glue, the task prompt, and some scissors; an additional copy of the diner plan may be useful.

MO'S DINER

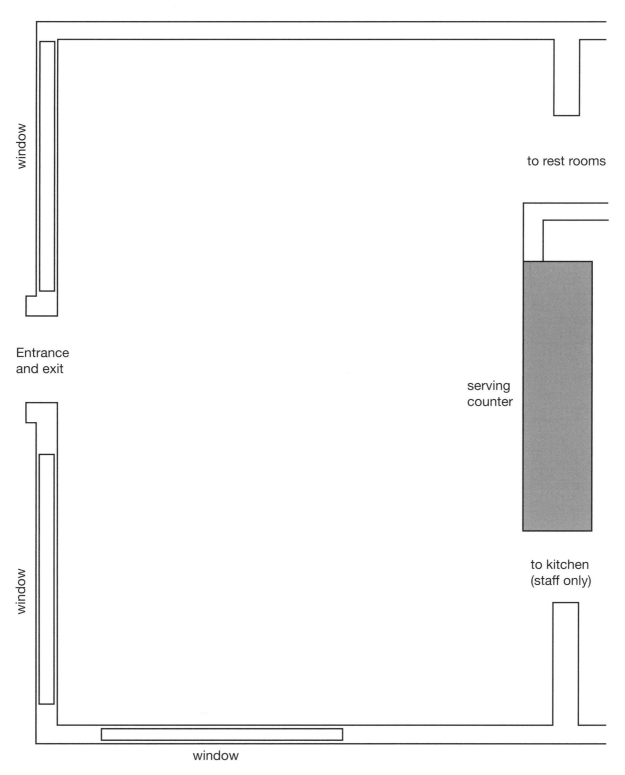

Floor plan to scale of 1 inch represents 4 feet

Student Work

Sample Response: Mo's Diner

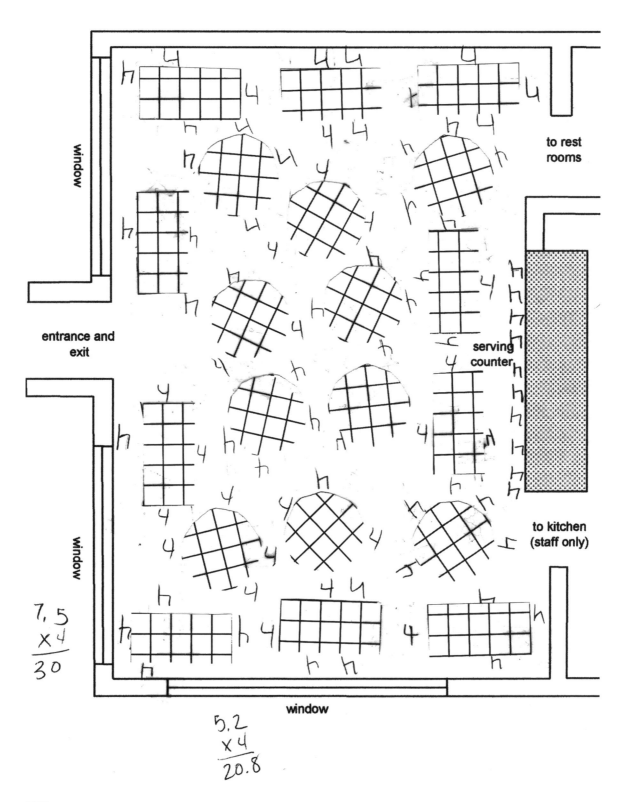

Sample Response: Mo's Diner

There are 10 Rectangular tables in the Diner. The 10 tables cost a total of 5699.90. To get the total cost I took the number of tables (10) times the cost per table ($569.99) 10 × 569.99 = $5699.90

There are 10 Round tables in the diner. Each round table cost $449.99 (per table). To find the total cost of the tabels I take the price for one table ($449.99) and times it by the number of tables I have (10) 449.99 × 10 = $4499.90

In this diner there are 4 chairs per table, 20 tables, and 10 chairs at the serving counter. The total number of chairs in this diner are 90. I got the number 90 from taking the number of chairs at all of the tables (80) and adding them with the number of chairs at the serving counter (10) 80 + 10 = 90 chairs total. 90 chairs times the price of one chair equals the total price. one chair cost $199.99.   90 × 199.99 = 17999.10

The total cost of how much all the furniture in the diner will cost is $28198.90. I figured it out by taking the total cost of Rectangular tables ($5699.90), Round tables ($4499.90), and the total cost of the chairs ($17999.10) and adding them together.

> **Standards:** Understand measurable attributes of objects and the units, systems, and processes of measurement; apply appropriate techniques, tools, and formulas to determine measurements

> **Expectations:** Understand both metric and customary systems of measurement; use common benchmarks to select appropriate methods for estimating measurements; select and apply techniques and tools to accurately find length, area, volume, and angle measures to appropriate levels of precision

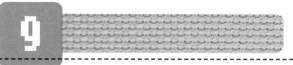

Randy drove 289 miles in 6 hours and then calculated his average rate of speed in miles per hour. Which of the following methods can he use to estimate whether his answer was reasonable?

Method 1: Multiply the rate by 6, and compare the product with 289.
Method 2: Divide 300 by the rate, and compare the quotient with 6.
Method 3: Divide the rate by 6, and compare the quotient with 289.

a. Methods 1 and 2 only
b. Methods 1 and 3 only
c. Methods 2 and 3 only
d. Methods 1, 2, and 3

> **Source:** Nova Scotia Department of Education Assessment Program
> **About the mathematics:** This item involves applying the formula $d = r \times t$.
> **Solution:** The correct response is a.
>   Method 1: The rate times the time must equal the distance.
>   Method 2: The distance divided by the rate must equal the time.

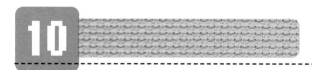

A liter of liquid will fill a cube measuring 10 cm on each edge.

a. What is the volume in cubic centimeters of a cube measuring 10 cm on each edge?
b. Use your answer from part (a) to find the number of milliliters in a liter.
c. What is the mass of a liter of water in grams?
d. A kilogram is 1000 grams. What is the mass of a liter of water in kilograms?

Source: *Middle Grades Maththematics*, "The STEM Project" (Evanston, Ill.: McDougal Littell, 1999, p. 440)

About the mathematics: This item requires understanding the metric system and using it as a tool for estimation.

Solution

a. The volume equals 10 cm × 10 cm × 10 cm, or 1000 cubic centimeters.
b. 1 liter equals 1000 milliliters.
c. 1 liter weighs 1000 grams.
d. 1 liter equals 1 kilogram.

Standards: Understand measurable attributes of objects and the units, systems, and processes of measurement; apply appropriate techniques, tools, and formulas to determine measurements

Expectations: Understand both metric and customary systems of measurement; select and apply techniques and tools to accurately find length, area, volume, and angle measures to appropriate levels of precision

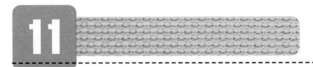

11

Some of the largest trees in the world are found on the west coast of North America. They include the grand sequoia trees, each of which pumps approximately 500 liters of water from its roots to its leaves every 12 hours. Approximately how much water will a sequoia tree pump from its roots to its leaves each year? Show your work.

Source: Adapted from Nova Scotia Department of Education Assessment Program

About the mathematics: Students apply common understandings of hours per day and days per year to solve this problem.

Solution: Pumping 24 hours a day means that a sequoia pumps 1000 liters per day.

$$1000 \text{ L} \times 365 = 365{,}000 \text{ L}$$

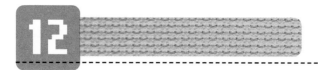

A mechanic needs to replace a machine belt that fits around the two wheels shown below. The radius of each wheel is 6 inches.

25 in.

1   How long must the belt be? Show all your work.
2. Justify the reasonableness of your answer.

Source: Connecticut Measurement and Geometry Practice Tasks (Hartford, Conn.: Connecticut Department of Education, 2002)

About the mathematics: To solve this problem, students must determine the circumference of a circle and understand the point of tangency of a circle.

Solution

1. $6\pi + 6 + 25 + 6 + 6\pi + 6 + 25 + 6 \approx 111.70$ inches.
2. $C = \pi d$; belt length = $12\pi + 74 \approx 111.68$ inches.

Two diameters plus the distance between the circles is $24 + 25 = 49$, and the belt must exceed twice that distance, which is 98 inches; so the answer is reasonable.

Rubric (5 possible points)

2 points: Arrives at correct solution of 111.70 inches (accept rounded answers)

1 point: Finds circumference of circle (e.g., 12 × 3.14)

1 point: Finds additional lengths from radius (e.g., 12 + 12)

1 point: Finds lengths of upper and lower portion of belt (25 + 25)

Student Work

Student Response A

This response indicates an understanding that pi would be involved; however, the student did not really show a complete understanding of the problem.

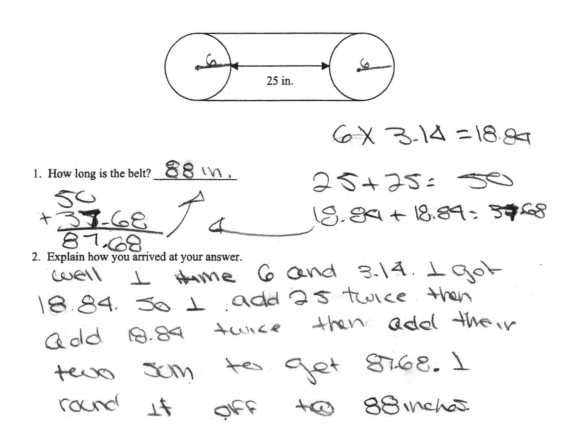

1. How long is the belt? __88 in.__

$$6 \times 3.14 = 18.84$$

$$25 + 25 = 50$$

$$18.84 + 18.84 = 37.68$$

$$\begin{array}{r} 50 \\ + 37.68 \\ \hline 87.68 \end{array}$$

2. Explain how you arrived at your answer.

well I time 6 and 3.14. I got 18.84. So I add 25 twice then add 18.84 twice then add their two sum to get 87.68. I round it off to 88 inches.

Student Response B

This response indicates a good understanding of the problem and the solution.

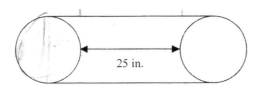

1. How long is the belt? ~~mmmmmm~~ 111.68 in

2. Explain how you arrived at your answer.

I first doubled the radius (6) to get the diameter. Then I multiplyed that by pie to get the circumerence of 1 circle. Since on each circle only half has a belt I only did it once. I added 25 plus the radios of the two circles(12). than I doubled that because there are two sides.

Student Response C

This response indicates that the solution is to sum two circumferences and add twice the distance between the circles.

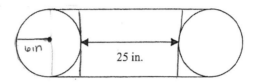

1. How long is the belt? __125.36 in__

2. Explain how you arrived at your answer.

The length from one cirle to the other is 25 in.

there are two sides 25in + 25in = 50in

the radius is 6 in, 4 equal radius each 90°

$\frac{1}{4}$ of the radius is 6 in so it would be

6·4 which equal 24in time it by two

because there are 2 circles and it would

be 48in.

$\pi = 3.14$

$\begin{array}{r} 48\text{in} \\ + 50\text{in} \\ \hline 98\text{in}. \end{array}$

$\begin{array}{r} 3.14 \\ \times\ 12 \\ \hline 628 \\ 314 \\ \hline 37.68 \end{array}$

$\begin{array}{r} 37.68 \\ \times\ \ 2 \\ \hline 75.36 \end{array}$

$C = \pi r 2$
$C = \pi (6in) r$
$C = \pi (12)$

$C = 3.14 \cdot 12$
$C = 37.68$

$\begin{array}{r} 75.36 \\ + 50.00 \\ \hline 125.36 \end{array}$

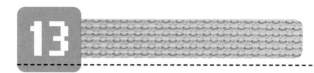

You are planning to tile the floor of a rectangular room that is $15\frac{1}{2}$ feet by $18\frac{1}{2}$ feet in size, using square tiles. The tiles measure 12 inches on each side and are sold in boxes of 25. How many boxes of tiles do you need to complete the job?

a. 9
b. 12
c. 34
d. 59

About the mathematics: This item requires students to use the formula for determining the area of a rectangle, and to understand that since each tile is 1 square foot, the solution simply involves dividing by the number of tiles per box.

Solution: The correct response is b.

$$15.5 \times 18.5 = 286.75 \text{ sq. ft.,}$$
$$\frac{286.75}{25} = 11.47;$$

so 12 boxes are needed.

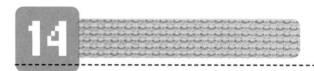

The figure on the following page shows a square inscribed in a larger square. What is the area of the smaller square inscribed in the larger square? Show all your work.

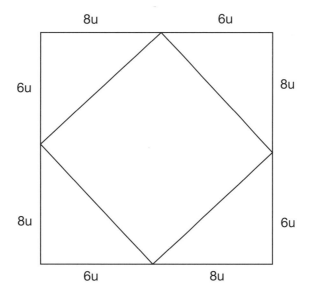

a. 36 sq. units
b. 64 sq. units
c. 100 sq. units
d. 196 sq. units

Source: Adapted from Nova Scotia Department of Education Assessment Program

About the mathematics: For this item, students find the area of an inscribed square by using the Pythagorean theorem to find the hypotenuse of a triangle given two side lengths.

Solution: The correct solution is c.

$$\left(6u\right)^2 + \left(8u\right)^2 = 36u + 64u^2 = 100u^2$$

$$\sqrt{100u^2} = 10u$$

$$\left(10u\right)^2 = 100u^2$$

Student Work

Student Response

In this response, the student has applied the rule of Pythagoras correctly.

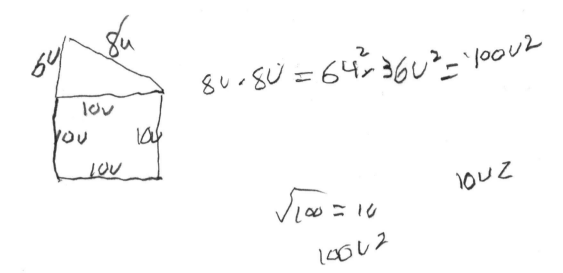

15

A circular pond is 17.5 meters in diameter and is surrounded by a path that is 2 meters wide.

a. Explain how you would determine the area of the path.

b. Find that area.

About the mathematics: This item involves applying the formula for the area of a circle in context and communicating a problem-solving plan in writing.

Solution: Determine the area of the outer circle (the path plus the pond), and subtract the area of the inner circle (the pond). The pond has a diameter of 17.5 meters; therefore, the radius of the pond is 17.5 ÷ 2 = 8.75 meters. Because the path is 2 meters wide, the radius of the pond plus the path is 10.75 meters. The area of the pond is $(8.75)^2 \times \pi = 76.56\pi$ square meters, and the area of the path-plus-pond region is $(10.75)2 \times \pi = 115.56\pi$ square meters. The area of the path is 115.56≠ – 76.56π, or 39 square meters.

Rubric (4 possible points)

4 points: Gives an explanation of the plan and the correct numerical solution

3 points: Gives an explanation of the plan but makes a small error in calculation

2 points: Gives an explanation but uses an outer diameter of 19.5 meters

1 point: Gives the correct solution but no explanation

0 points: Gives neither an explanation of the plan nor a solution

16

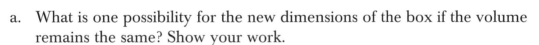

A cereal company decided to increase the height of its boxes by 20 percent. If the volume remains the same, what could be the new dimensions of the width and length?

The original dimensions are as follows:

length = 20 cm
height = 40 cm
width = 30 cm

a. What is one possibility for the new dimensions of the box if the volume remains the same? Show your work.

b. Does it make sense to have a cereal box with these dimensions? Justify your answer.

Source: St. Francis Xavier University, Nova Scotia, http://www.stfx.ca/special/mathproblems (accessed March 14, 2005)

About the mathematics: This item involves finding the percent of a number, calculating the volume of a rectangular prism, and estimating the size of a normal storage shelf.

Solution: The volume is 24,000 cubic centimeters; the new height is 48 centimeters; possible dimensions for the widths and lengths are 20 and 25 centimeters, 10 and 50 centimeters, 12.5 and 40 centimeters. Students should consider whether the cereal box would fit on an average-sized shelf.

Rubric (3 possible points)

1 point: Correctly calculates the new height

1 point: Arrives at a pair of dimensions to be used with the height

1 point: Determines and justifies whether the new dimensions are reasonable

–or–

0.5 points: States only that the new dimensions are not reasonable

Student Work

Student Response A

In this response, the cereal box height would be 30 percent taller, hence, 52 inches.

$$20 \cdot 30 \cdot 40 = 24000$$
$$20 \cdot 52 \cdot W = 24000$$
$$1040 \times W = 24000$$
$$W = 23.076$$

Student Response B

In this response the student increases the height by 30 percent, but then to compensate, notes that the width must be reduced by 30 percent.

One possibility for the new dimension of the box if the volume remains the same is that the length would be the same 20 cm, the height would be 30% tall which is 52 cm and the width would be 30% smaller which is 21 cm.

Standards: Understand measurable attributes of objects and the units, systems, and processes of measurement; apply appropriate techniques, tools, and formulas to determine measurements

Expectations: Understand both metric and customary systems of measurement; develop and use formulas to determine the circumference of circles and the area of triangles, parallelograms, trapezoids, and circles and develop strategies to find the area of more-complex shapes.

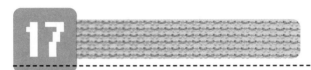

Four identical triangles are arranged inside a rectangle as shown. The figure is not drawn to scale.

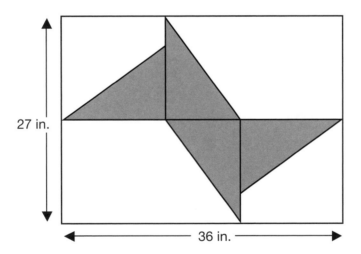

What is the area of one of the triangles?

Source: Adapted task from http://www.worldclassarena.org/v5/example_ questions.htm; reproduced by permission of nferNelson, The Chiswick Centre, 414 Chiswick High Road, London, W4 5TF, copyright © QCA

About the mathematics: This item requires the ability to identify the base and height of triangles regardless of their orientation and to use those measurements to find the area of one of the triangles.

Solution

$$2h = 27 \text{ in.};$$

therefore,

$$h = 13.5 \text{ in.},$$
$$h + b + h = 36,$$
$$2h + b = 36,$$
$$27 + b = 36;$$

therefore,

$$b = 9 \text{ in.}$$
$$A = \tfrac{1}{2} bh = \tfrac{1}{2} \times 13.5 \times 9 = 60.75 \text{ sq. in.}$$

Student Work

Student Response A

This response indicates confusion about the length of the base of a triangle. Although the formula for the area of a triangle appears, the student's squaring, rather than halving, of 13.5 shows a misconception about how to solve the problem.

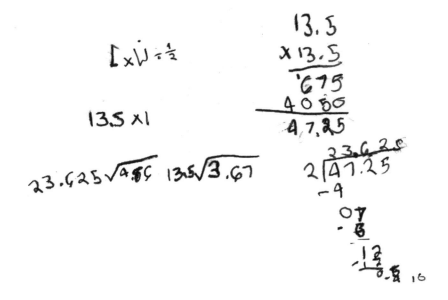

Student Response B

This response indicates that the base of the triangle would be $^1/_4$ of the length of the rectangle. How this relationship was derived is not clear. Perhaps it was an estimate.

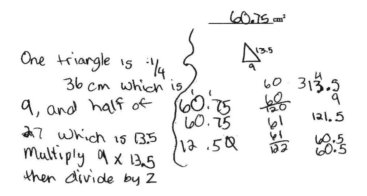

18

Four identical triangles are arranged inside a rectangle, as shown. The figure is not drawn to scale.

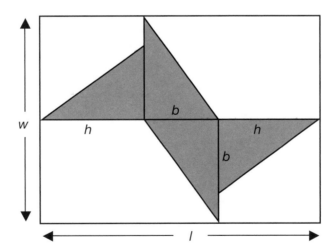

1. Find the area of the rectangle in terms b and h. Show all your work.
2. Explain how you would find the perimeter of the rectangle in terms of b and h.

Source: Adapted task from http://www.worldclassarena.org/v5/example_ questions.htm; reproduced by permission of nferNelson, The Chiswick Centre, 414 Chiswick High Road, London, W4 5TF, © QCA

About the mathematics: This item requires the ability to identify the base and height of triangles regardless of their orientation and to use those measurements to find the area and perimeter of those triangles using the general values of *b* and *h*.

Solution

$A = (2h + b)(2h)$. Since one length is $(h + b + h)$ and one width is $2h$, the perimeter, *P*, is $2(2h + b) + 2(2h)$, or $8h + 2b$.

Rubric (2 possible points)

2 points: Both correct area and correct perimeter found, with work shown

1 point: Either correct area or correct perimeter found

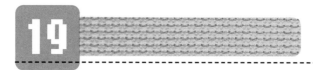

19

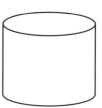

A farm has a vertical cylindrical oil tank that has an inside diameter of 2.5 feet. The depth of the oil in the tank is 2 feet. If 1 cubic foot of space holds 7.48 gallons, about how many gallons of oil are in the tank?

a) 59 gallons
b) 75 gallons
c) 230 gallons
d) 294 gallons

Source: Adapted from WorkKeys Assessments, Applied Mathematics, © 2003 by ACT, http://www.act.org/workkeys/assess/math/sample7.html (accessed March 14, 2005)

About the mathematics: This item requires the ability to find the volume of a cylinder and to determine what volume of liquid fills in a certain portion of that cylinder.

Solution: The correct response is b. Calculate $3.14 \times 1.25^2 \times 2 = 9.8125 \approx$ 10 cubic feet; $10 \times 7.48 \approx 75$ gallons.

Student Work

Student Response

In this example, the student attempted to draw a model of the oil tank. The drawing indicates 2 feet as being the height of the cylinder instead of the height of the oil in the tank. This student chose the correct response, but the work does not indicate how the student reasoned about all the factors.

Standards: Understand measurable attributes of objects and the units, systems, and processes of measurement; apply appropriate techniques, tools, and formulas to determine measurements

Expectations: Understand both metric and customary systems of measurement; develop strategies to determine the surface area and volume of selected prisms, pyramids, and cylinders

---

20

The polygon *ABCD* shown is a square. Point *E* is the midpoint of *AB*, and point *F* is the midpoint of *BC*. Point *D* is connected to each of the points *E*, *B*, and *F*, thus dividing the square into four triangles: △*DAE*, △*DEB*, △*DBF*, and △*DFC*.

1. How do the areas of the triangles compare?
2. Show that your conclusions hold for any square.

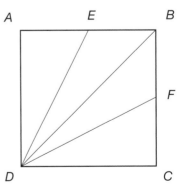

Source: Adapted from *Do Bees Build It Best?* Interactive Mathematics Program, Year 2 (Emeryville, Calif.: Key Curriculum Press, 1989, p. 224); 1150 65th Street, Emeryville, CA 94608; 1-800-995-MATH; www.keypress.com (accessed March 15, 2005)

About the mathematics: This item requires the ability to determine the relationship among four triangles constructed by drawing a diagonal and by drawing midpoints.

Solution

The area of each triangle is equal to

$$\tfrac{1}{2}\text{ base} \times \text{height.}$$

If *a* is the length (height) of a side of the square, the base of each triangle is $\tfrac{1}{2}a$. The area of each triangle is

$$\tfrac{1}{2}a \times \tfrac{1}{2}a = \tfrac{1}{4}a^2,$$

or one-fourth the area of the whole square.

Rubric

Level 4: The student's work meets the essential demands of the task.

- Student determines the area of each triangle in general terms either verbally or symbolically.
- Student provides a complete justification for why this approach works for any square.

Level 3: The student's work needs to be revised.

- Student correctly determines or computes the areas of the triangles by substituting specific values for a side length of the square.
- Student may pose a symmetry argument but does not justify the result.

Level 2: The student needs some instruction.

- Student attends to segments or angles that are congruent but does not relate statements to the base or height of the triangle.
- Student might verify two of the triangles as being congruent but may believe that the other two are of a different size.

Level 1: The student needs significant instruction.

- Student does not engage with the task, or states the triangles are all equal without supplying a reasonable supporting argument.

Student Work

Student Response A

The student's first assumption, which is incorrect, is that the line segments are equal. The second conclusion is that the triangles are equal in pairs. The student makes no attempt to answer the second question concerning the conclusion in all squares.

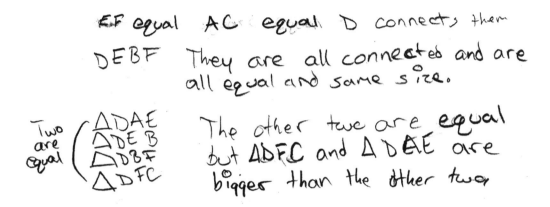

Student Response B

This student makes no to attempt to make any conclusion for all the triangles. The conclusion here is that since the points are all midpoints, the triangles must all be equal in area.

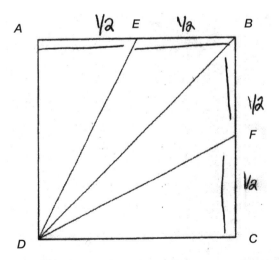

Student Response C

This student's approach is to model the problem by assigning 10 to the length of the side of the square. Although the student indicates on the paper that the area of a triangle is $1/2$ the base times the altitude, that fact is ignored, and the student arrives at the conclusion that the area of each triangle is 50 square units. The student does indicate, however, that the area of the square is 100 square units. The student makes no attempt to generalize the solution.

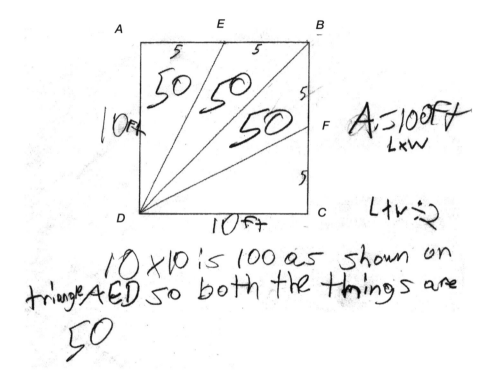

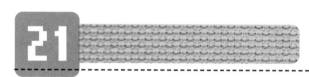

21

Two circles have radii of 2 cm and 6 cm, respectively. How many times larger in area is the larger circle than the smaller circle?

a. 3 times as large
b. 4 times as large
c. 9 times as large
d. 12 times as large

About the mathematics: This item involves using proportional reasoning to determine a scale factor.

Solution: The correct response is c.

$$36\pi \div 4\pi = 9$$

Student Work

Student Response

The work shows the student's ability to find the areas of both circles and to make comparisons.

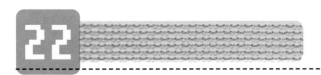

$$3.14 \times 2 \times 2 = 12.56$$
$$3.14 \times 6 \times 6 = 113.04$$
$$\frac{113.04}{12.56} = 9$$

22

A company manufactures large boxes of one size. If the company decides to make a new box with twice the length, width, and height of the old box, how will the volume of the new box compare with the volume of the old box?

a. It will be two times greater.
b. It will be four times greater.
c. It will be eight times greater.
d. It will be sixteen times greater.

Source: Nova Scotia Department of Education Assessment Program

About the mathematics: Proportional reasoning used in determining scale a factor.

Solution: The correct response is c.

$$2l \times 2w \times 2h = 8lwh$$

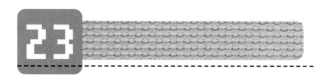

Jean had a large rectangular piece of fabric. She cut a piece of the fabric that is half as wide and half as long as the original piece. Show how the area of the piece she cut compares with the area of the original piece?

Source: Adapted from Nova Scotia Department of Education Assessment Program

About the mathematics: For this item, students use proportional reasoning to determine a scale factor

Solution: The area of the cut piece of fabric is one-fourth the area of the original piece of fabric.

$$^1/_2 w \times ^1/_2 l = ^1/_4 wl$$

Student Work

Student Response A

The work indicates good understanding of problem.

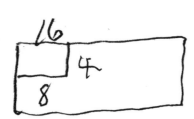

Student Response B

The student neglected to determine the ratio of the two areas, which were found correctly. The use of the variable may have caused confusion.

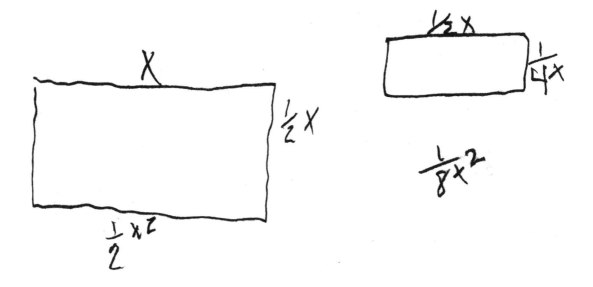

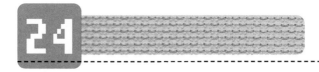

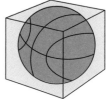

24

Slam Dunk Sporting Goods packages its basketballs in cubic boxes with 1-foot edges:

a. Slam Dunk ships basketballs from its factory to stores all over the country. To ship the balls, the company packs 12 basketballs (in their boxes) into a large rectangular shipping box. Find the dimensions of every possible shipping box into which the boxes of balls would fit tightly without additional packing.
b. Find the surface area of each shipping box you found in part a.
c. Slam Dunk uses the shipping box that requires the least material. Which shipping box does the company use?
d. Slam Dunk decides to ship basketballs in boxes of 24. The company wants to use the shipping box that requires the least material. Find the dimensions of the box that should be used. How much more packaging material is needed to ship 24 boxed balls than to ship 12 boxed balls?

Source: Adapted from *Connected Mathematics Project,* "Filling and Wrapping" (Upper Saddle River, N.J.: Pearson Prentice Hall, © 1996 Michigan State University, p. 83, problem 1)

About the mathematics: This item requires the ability to determine the whole-number factors of 12 and the surface area of a rectangular prism.

Solution

a) 1 foot by 1 foot by 12 feet, 1 foot by 2 feet by 6 feet, 1 foot by 3 feet by 4 feet, and 2 feet by 2 feet by 3 feet

b) The surface areas of the boxes above are 50 square feet, 40 square feet, 38 square feet, and 32 square feet, respectively.

c) The 2-foot-by-2-foot-by-3-foot box

d) Any of the arrangements in part (a) with one of its dimensions doubled will result in a box that would hold 24 basketballs. The one that uses the least amount of material is a 2-foot-by-3-foot-by-4-foot box, requiring 52 square feet of material, which is 20 square feet more than the box that requires the least amount of material to hold 12 basketballs.

Rubric (12 possible points)

a) *4 points:* Identifies each of four boxes with whole-number dimensions yielding a volume of 12 cubic feet

b) *4 points:* Finds correct surface area for each box in part (a)

c) *1 point:* Correctly identifies the box with the least surface area

d) *3 points:* Correctly finds the following (1 point each):
   - A box that ships 24 balls
   - The box with least surface area
   - The difference in surface area between boxes that require the least amount of material to hold 24 and 12 basketballs, respectively

Student Work

Student Response

In this response, the student seems to deal only with boxes 1 foot wide. The work in part b reveals confusion between volume and surface area. The work in part c indicates a misconception that the boxes are made of wood and a lack of understanding of surface area. Part d also indicates a lack of understanding of surface area.

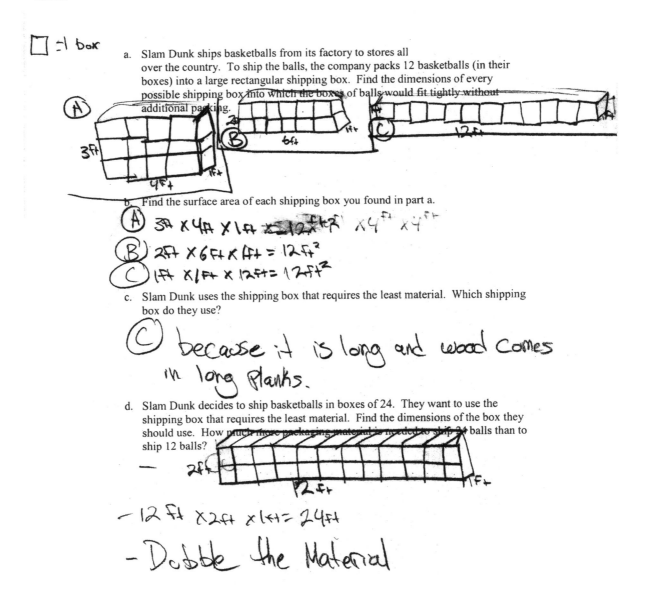

□ =1 box

a. Slam Dunk ships basketballs from its factory to stores all over the country. To ship the balls, the company packs 12 basketballs (in their boxes) into a large rectangular shipping box. Find the dimensions of every possible shipping box into which the boxes of balls would fit tightly without additional packing.

b. Find the surface area of each shipping box you found in part a.

(A) 3ft × 4ft × 1ft = 12ft² × 4ft × 4ft

(B) 2ft × 6ft × 1ft = 12ft²

(C) 1ft × 1ft × 12ft = 12ft²

c. Slam Dunk uses the shipping box that requires the least material. Which shipping box do they use?

(C) because it is long and wood comes in long planks.

d. Slam Dunk decides to ship basketballs in boxes of 24. They want to use the shipping box that requires the least material. Find the dimensions of the box they should use. How much more packaging material is needed to ship 24 balls than to ship 12 balls?

– 12 ft × 2ft × 1ft = 24ft

– Double the Material

Standards: Understand measurable attributes of objects and the units, systems, and processes of measurement; apply appropriate techniques, tools, and formulas to determine measurements

Expectations: Understand both metric and customary systems of measurement; select and apply techniques and tools to accurately find length, area, volume, and angle measures to appropriate levels of precision; solve simple problems involving rates and derived measurements for such attributes as velocity and density

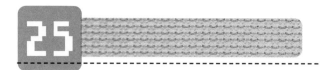

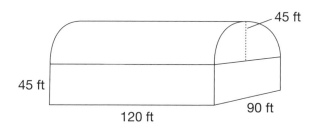

A small aircraft hangar looks like a rectangular prism topped by half of a cylinder. A painter has been hired to paint the outside of the building. A heating contractor has been hired to install a heating unit inside the building.

a. If a gallon of paint covers 500 square ft. How much paint will the painter need? Show all your work. Justify your answer.
b. Calculate the space that needs to be heated? Show all your work. Justify your answer.

Source: From *MathScapes: Shapes and Spaces* (Creative Publications, 1998; available from Glencoe/McGraw-Hill)

About the mathematics: This item requires students to determine the surface area and volume of an irregular geometric object.

a) The surface area of the sides of the building (excluding base) is
$$2(45 \times 120) + 2(45 \times 90) = 18{,}900$$
square feet. The surface area of the two half-circles, or one whole circle, is
$$2 \times \tfrac{1}{2}\pi \times 45 \times 45 = (45)^2\pi \approx 6358.50$$
square feet. The remaining surface area of the half-cylinder equals
$$120 \times \tfrac{1}{2}\pi \times 90 = 5400\pi \approx 16956$$
square feet. The total surface area is
$$18{,}900 + 6358.50 + 16956 \approx 42{,}214.5$$
square feet.

b) The volume of the rectangular prism is
$$45 \times 120 \times 90 = 486{,}000$$
cubic feet. The volume of the half-cylinder is
$$\tfrac{1}{2}(\pi \times 45 \times 45 \times 120) \approx \tfrac{1}{2}(763{,}020) \approx 381{,}510$$
cubic feet. The total volume of the building is approximately 867,000 cubic feet.

Rubric

Level 4: Goes beyond expectations

- Shows a well-developed ability to find the surface area and volume of a prism using the correct formulas or another acceptable method
- Shows a well-developed ability to find the surface area and volume of a cylinder using the correct formulas or another acceptable method
- Includes correct calculations of the surface area and volume of a compound figure
- Includes a well-defined solution demonstrating excellent understanding of the questions

Level 3: Meets all expectations

- Shows ability to find the surface area and volume of a prism using the correct formulas or another acceptable method
- Shows ability to find the surface area and volume of a cylinder using the correct formulas or another acceptable method
- Includes correct calculations of the surface area and volume of a compound figure
- Includes an acceptable solution demonstrating understanding of the questions

Level 2: Meets some expectations
- Shows some ability to find the surface area and volume of a prism using the correct formulas or another acceptable method
- Shows some ability to find the surface area and volume of a cylinder using the correct formulas or another acceptable method
- Makes errors in calculations of the surface area and volume of a compound figure
- Arrives at an unacceptable solution demonstrating limited understanding of the questions

Level 1: Falls below expectations
- Shows inability to find the surface area and volume of a prism using the correct formulas or another acceptable method
- Shows inability to find the surface area and volume of a cylinder using the correct formulas or another acceptable method
- Makes an unsuccessful attempt to find the surface area and volume of a compound figure
- Shows little or no understanding of the questions

Student Work

Student Response

The work contains an error is in determining the area of the circular ends of the roof.

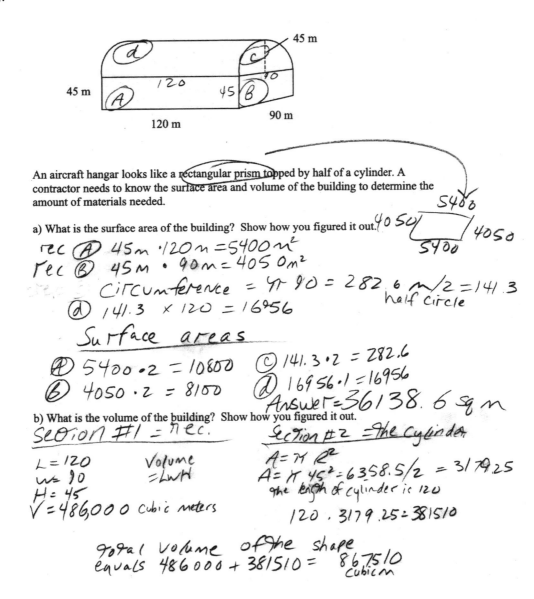

An aircraft hangar looks like a rectangular prism topped by half of a cylinder. A contractor needs to know the surface area and volume of the building to determine the amount of materials needed.

a) What is the surface area of the building? Show how you figured it out. 4050 5400 4050 5400

rec Ⓐ 45m · 120m = 5400 m²
rec Ⓑ 45m · 90m = 4050 m²
       Circumference = 4π 90 = 282.6 m/2 = 141.3
                                                half circle
Ⓓ 141.3 × 120 = 16956

Surface areas
Ⓐ 5400 · 2 = 10800   Ⓒ 141.3 · 2 = 282.6
Ⓑ 4050 · 2 = 8100   Ⓓ 16956 · 1 = 16956
                           Answer = 36138.6 sq m

b) What is the volume of the building? Show how you figured it out.
Section #1 = rec.   Section #2 = the Cylinder

L = 120   Volume   A = π R²
w = 90   = LwH   A = π 45² = 6358.5/2 = 3179.25
H = 45   the length of cylinder is 120
V = 486,000 cubic meters
                               120 · 3179.25 = 381510

        Total volume of the shape
        equals 486000 + 381510 = 867510
                                   cubic m

Standards: Understand measurable attributes of objects and the units, systems, and processes of measurement; apply appropriate techniques, tools, and formulas to determine measurements

Expectations: Understand both metric and customary systems of measurement; select and apply techniques and tools to accurately find length, area, volume, and angle measures to appropriate levels of precision; solve problems involving scale factors, using ratio and proportion

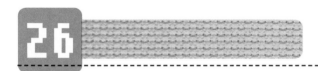

Orlando bought a plot of land measuring 1 square mile, on which he plans to grow blueberries. He hires 346 people to plant seedlings. It takes about 20 seconds to plant each seedling, and each seedling requires 1.5 square feet of area.

a. How many square feet of land can the 346 workers plant each minute?
b. How much time will it take to plant the entire plot of land? Explain your reasoning.
c. Orlando decides to plant a different type of blueberry that requires only 1.2 square feet of land per seedling. How much time will it take the workers to plant these seedlings?

> **Source:** Adapted from *Connected Mathematics,* "Data around Us" (Upper Saddle River, N.J.: Prentice Hall, 1996, p. 132, item1)
>
> **About the mathematics:** This item involves finding rate of time to complete a task given a measure of land.
>
> **Solution**
>   a. 1557 square feet
>   b. A square mile contains (5280 feet/mile)^2, or 27,878,400 square feet. The workers will take 27,878,400/1557, or about 17,905, minutes, which equals about 298 hours, to plant the entire plot.
>   c. About 373 hours

A map uses a scale of 1 cm = $5\,^1/_2$ mi. In actual distance, the entrances to two parks 24 3/4 are miles apart. How far apart are they on the map?

a. $30\,^1/_2$ cm
b. $9\,^1/_4$ cm
c. $5\,^1/_2$ cm
d. $4\,^1/_2$ cm

Source: Adapted from Nova Scotia Department of Education Assessment Program

About the mathematics: For this item, students apply metric measure to a contextual problem involving division of mixed numbers.

Solution: The correct solution is d: $24\,^3/_4 \div 5\,^1/_2 = 4\,^1/_2$ cm.

Student Work

Student Response

The work indicates an understanding of the conversion process; however, it demonstrates a misconception in the division of mixed numbers.

$$24\,^3/_4 = \frac{96 \times 3}{4}$$

$$\frac{\frac{99}{4}}{\frac{11}{2}} = \frac{9}{\cancel{99}} \times \cancel{\frac{2}{4}}\Big/ \cancel{\frac{1}{4}} = \frac{9}{2} = 4\frac{1}{2}$$

Standards: Understand measurable attributes of objects and the units, systems, and processes of measurement; apply appropriate techniques, tools, and formulas to determine measurements

Expectations: Understand both metric and customary systems of measurement; understand, select, and use units of appropriate size and type to measure angles, perimeter, area, surface area, and volume; solve problems involving scale factors, using ratio and proportion

Use the diagram to answer the question.

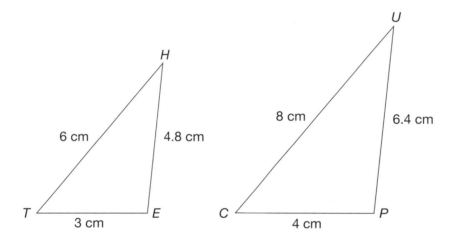

Complete the following:
a. Give the ratio, in lowest terms, of each set of corresponding sides.
b. Are the triangles similar? Why or why not?

Source: Nova Scotia Department of Education Assessment Program

About the mathematics: This item requires students to understand that if all three pairs of corresponding side lengths are in the same proportion, the triangles are similar.

Solution

a. $^6/_8 = ^3/_4$; $^3/_4$; $^{4.8}/_{6.4} = ^6/_8$, and $^6/_8 = ^3/_4$.

b. If all three pairs of corresponding side lengths of two triangles are in the same proportion, then the triangles are similar.

Rubric (2 possible points)

2 points: Gives correct ratios in simplified terms and a correct explanation about similarity

1 point: Gives correct ratios in simplified terms but an incorrect explanation of what makes triangles similar

–or–

Does not give ratios in lowest terms but supplies correct explanation of what makes triangles similar

0 points: Gives incorrect ratios and explanation

28

The following rectangle has a base of 5 meters and an area of 30 square meters.

$A = 30 \text{ m}^2$

5 m

If the rectangle is enlarged by a scale factor of 3 find the following:

a. Length of the new base
b. Length of the new height
c. Area of the enlarged figure

About the mathematics: This item involves applying scale factors to linear and spatial measure.

Solution

a. 5 m × 3 = 15 m.
b. 6 m × 3 = 18 m.
c. 30 m² × 9 = 270 m².

5 Data Analysis and Probability

PRIOR to entering the middle grades, students are expected to have had various experiences collecting, organizing, and representing data sets. Those past experiences will have given students ample opportunities to develop fluency with tables, line plots, bar graphs, and line graphs. Most students will have become familiar with finding measures of center and spread, including mean, median, mode, and range. "Although the mean often quickly becomes a method of choice for students when summarizing a data set, their knack for computing the mean does not necessarily correspond to a solid understanding of its meaning or purpose" (McClain 1999, cited in NCTM 2000, pp. 250–51). Students in the middle grades are expected to develop a conceptual understanding of mean as process of "evening out," or "balancing," a data set and of median as identifying the middle of the data set.

In the middle grades, students are expected to engage in examining populations and samples. They should be involved in answering more-complex questions concerning relationships among populations or samples and those about relationships between two variables within one population or sample (NCTM 2000, p. 249). Students are expected to add to their repertoire of representations for data sets, understanding how each representation portrays a data set. Box plots, scatterplots, histograms, and stem-and-leaf plots should be used appropriately, depending on the data sets being examined.

Students in the middle grades should also have numerous opportunities to engage in probabilisitic thinking about simple situations from which students can develop notions of chance (NCTM 2000, p. 253). Although the computation of probabilities can appear to be simple work with fractions, students must grapple with many conceptual challenges to understand probability. Students are helped by making predictions and then comparing the predictions with actual outcomes (p. 254).

A crucial point for educators to understand is that students need multiple opportunities over time and across contexts to demonstrate fluency, flexibility, and a deep understanding of statistical ideas.

This set of questions is designed to provide examples for assessing understanding of central tendencies, dispersions, and probabilities beyond the ability to compute. The items are presented in a variety of formats, including problem situations, tables, and graphs. Examples of types of questions that teachers can ask students to move them from an operational understanding to an ability to reason flexibly within and among the different measures are also presented. The set does not constitute complete coverage of the data analysis and probability topics in the middle-grades curriculum.

Data Analysis and Probability Assessment Items

Standard: Formulate questions that can be addressed with data and collect, organize, and display relevant data to answer them

Expectation: Formulate questions, design studies, and collect data about a characteristic shared by two populations or different characteristics within one population; select, create, and use appropriate graphical representations of data, including histograms, box plots, and scatterplots

Standard: Select the appropriate statistical methods to analyze data

Expectation: Discuss and understand the correspondence between data sets and their graphical representations, especially histograms, stem-and-leaf plots, box plots, and scatterplots

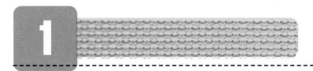

A manufacturer embroiders names on jackets. Each letter costs $0.50 to embroider. The company wants to make 100% profit on embroidery but wants to charge a flat fee per name rather than different amounts for different names. Most people want just their first names on their jacket. How should the company solve this problem? Devise a strategy to help the company, and then write a letter to the company showing the statistical methods that led to your suggested solution.

Source: Adapted from Charlotte Danielson, *Performance Tasks and Rubrics: Middle School Mathematics* (Larchmont, N.Y.: Eye on Education, 1997, p. 131)

Solution: Solutions will vary depending on the approach used by the student. Two possible approaches are the following:

a. Determining the number of letters in the names of students in the student's class, and, realizing that this sample may not be typical, extrapolating from that number to the rest of the school and larger universe of young people, then drawing a conclusion from the data.

b. Acquiring a listing of all students in the school, sampling a subset chosen randomly, counting the number of letters in the first names of the students in the sample, graphing the results, and examining them to draw a conclusion

Rubric (4 possible points)

4 points: The work is well organized; accurate; makes a good display of data; uses a large sample; and contains a well-thought-out argument, presentation, and conclusion.

3 points: The work is adequately organized; accurate; makes a good display of data; and draws valid, clear conclusions but uses a sample that is too small.

2 points: The work is poorly organized; contains some minor calculation errors or errors in data display; uses a sample that is too small; and draws valid conclusions but through methods that are not clear.

1 point: The work evidences no idea of how to determine average number of letters; the data are poorly organized; and the conclusions are invalid and poorly explained.

0 points: No organized effort is discernable; no conclusion or an invalid conclusion is reached.

Standard: Select the appropriate statistical methods to analyze data

Expectation: Find, use, and interpret measures of center and spread, including mean and interquartile range

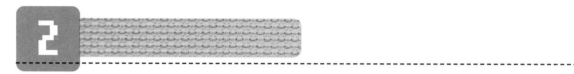

A newspaper surveyed some stores on the prices of all of their running shoes. The prices to the nearest dollar are displayed in the plot below.

```
 2 | 4 9
 3 | 2 5 5
 4 | 4
 5 | 2 2 4 8 9
 6 | 0 5 5 8 9
 7 | 0 4 4 9 9 9
 8 | 2 5 6
 9 | 4
10 | 5
11 | 0 9              9|4 = $94
```

A reporter writes, "Half of the running shoes in the stores surveyed are priced greater than or equal to _____." Which of the following numbers should the reporter use in the blank?

a. $82
b. $68
c. $67
d. $79

About the mathematics: This item involves the ability to determine the median of a data set.

Solution: The correct choice is b. Half the tennis shoes will be priced at least as much as the median, which is $68.

Distractors

$82 is the upper quartile.

$67 is the mean rounded to nearest dollar.

$79 is the mode.

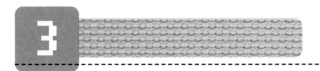

(Maximum Temperature, Average Precipitation) from 73 Major City Airports in the U.S.

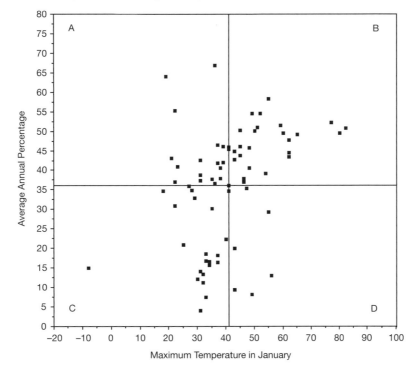

In this graph the vertical and horizontal lines intersect at the point (41, 36.2), where 41° F is the mean of the high temperatures and 36.2 inches is the mean of the average precipitation (either rain or snow) for the month of January. The two lines divide the graph into four parts, labeled A, B, C, D as indicated on the graph. In which of these four parts would you find an airport where the January high temperature was above the mean but the amount of precipitation was lower that the mean average temperature?

a. A

b. B

c. C

d. D

About the mathematics: This item requires students to interpret ordered pairs relative to the point (mean, mean).

Solution: The correct response is d.

Distractors

1. In part A the points are below the mean high temperature and above the average mean precipitation.
2. In part B the points are above the mean high temperature and above the average mean precipitation.
3. In part D the points are below the mean high temperature and below the average mean precipitation.

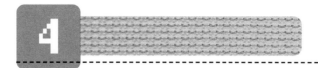

The box plot represents the test scores of students in a middle school mathematics class.

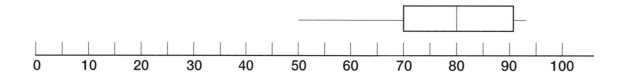

Which of the statements could be true of the test scores earned by the class?

a. Three-fourths of the class had a test score lower than 80.
b. Half of the class had a test score between 70 and 90.
c. Three-fourths of the class had test score 70 or above.
d. Half of the class had a test score between 70 and 80.

About the mathematics: Ability to use knowledge of box-plot construction in interpreting a related question.

Solution: Choice c is correct. Seventy-five percent of the scores in a box plot lie above the lower quartile.

Distractors

a. The median is approximately 80, identifying the middle value.
b. Half the scores are within the box, but the upper quartile is greater than 90.
d. One-fourth of the class scored between 70 and 80.

About the student work: From this question one can learn whether students know that each quartile of the box plot contains one-fourth of the data. The majority of the responses in the pilot group indicate that students did know this fact but had difficulty determining the upper-quartile value, or the upper end of the box. That fact led to many selections of choice b.

Student Work

Student Response A

This response shows a knowledge that each segment of the box plot represents one-fourth of the data points and thereby logically concludes the correct answer.

a) Three-fourths of the class had a test score lower than 80.
b) Half of the class had a test score between 70 and 90.
c) Three-fourths of the class had test score 70 or above.
d) Half of the class had a test score between 70 and 80.

The whisker on the left is. 1/4 of the class.
that's from 50 - 70.

Student Response B

This response indicates an understanding that each segment of the box plot represents one-fourth of the data points; the student appears to use the fractions ordinally to arrive at the correct response.

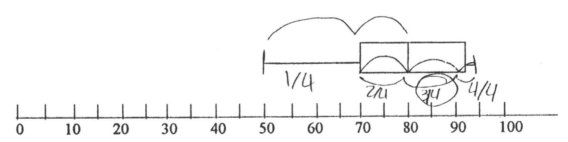

Which of the statements could be true of the test scores earned by the class?

a) Three-fourths of the class had a test score lower than 80.
b) Half of the class had a test score between 70 and 90.
c) Three-fourths of the class had test score 70 or above.

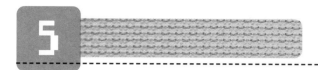

The following table lists the selling prices of 28 homes in Milwaukee, Wisconsin, listed in the *Milwaukee Journal Sentinel* on August 1, 2004. You work for a realtor, and are asked to report the "average" selling price of a home in a newspaper article. Would you use the mean or the median? Explain your reasoning.

$229,900	$264,500
$1,095,900	$99,000
$353,900	$242,000
$264,900	$449,900
$218,000	$219,900
$624,000	$84,900
$980,000	$112,900
$91,900	$189,900
$299,500	$544,900
$749,000	$220,000
$892,500	$149,900
$538,440	$219,900
$975,000	$389,450
$579,000	$544,900

About the mathematics: This item involves reasoning from data, and identifying a representative statistic.

Solution: The mean is $415,142.50, and the median is $282,200. The median price is more realistic because it gives the middle price, indicating that 50 percent of the houses cost more than $282,200 and 50 percent cost less. Students choosing the mean will not have accounted for the outlier in the data set.

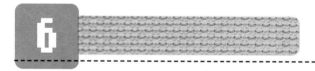

If the same number is added to *each* data point in a data set, which of the following statements will be true?

a. The range is increased by double the number added.
b. The median is unchanged.
c. The interquartile range is unchanged.
d. The mean is unchanged.

> **About the mathematics:** This item involves the ability to reason that adding a number to each data point translates the graph horizontally without affecting the range or the interquartile range.
>
> **Solution:** The answer is c. The values of the mean and median will increase by the value of the constant, but the range and interquartile range will remain unchanged.
>
> **Distractors**
>
> a. The graph will simply be translated right or left according to the value of the constant, so the range will remain unchanged.
> b. Since the graph will be translated, the median will have to change value.
> d. Since the graph will be translated, the mean will have to change value.

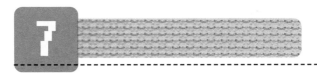

If *each* data point in a data set is multiplied by the same number, which of the following statements will be true?

a. The range is unaffected.
b. The median is doubled regardless of the number used.
c. The mean is equal to the constant times the original mean.
d. The interquartile range remains the same.

> **About the mathematics:** This item involves the ability to imagine the effect that multiplication will have on a graphical representation. (Dilation)
>
> **Solution:** The answer is c. All data points will increase by being multiplied by the constant, affecting all values but not their differences.
>
> **Distractors:**
>
> a. The range will be equal in length to the original multiplied by the constant.
> b. The median will be doubled only if the constant is 2 or –2.
> c. The interquartile range will be equal to the constant times the original.

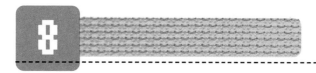

Pets of Seventh Graders

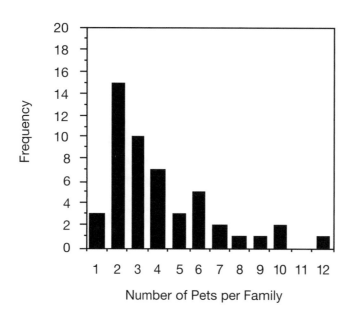

Number of Pets per Family

According to the graph, which of the following statements is true?

a. The median number of pets per family is more than the mean.
b. The mean number of pets per family is more than the median
c. The mean number of pets per family is 2.
d. Over half the families have at least 6 pets.

> **About the mathematics:** This item involves the ability to retrieve information from a graph and use it to solve a problem requiring multiple steps.

Solution: The correct choice is b. The graph in item 8 is skewed as compared with a bell-shaped graph (shown here):

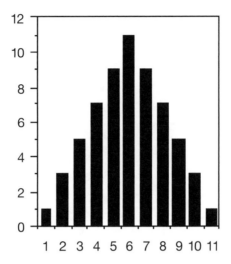

In a bell-shaped graph the mean and median are in the middle. In a skewed graph, as given in the example, the mean is affected by the greater spread of the values and is pulled away from the middle toward the tail. As a result, the mean will become larger or smaller than the median. In this example the mean is becoming larger than the median.

Arithmetically we could calculate the mean:

$$\frac{3\times1+15\times2+10\times3+7\times4+3\times5+5\times6+2\times7+1\times8+1\times9+2\times10+1\times12}{50} = 3.98 \approx 4$$

The median of the data, 3, can be deduced from the graph, or the student can order the data and count to the middle value.

Distractors
  a. The opposite is true.
  c. The mean is actually 4.
  d. One can observe from the graph that the number of families that have 6 pets is 5, which, reasoned on the basis of the other information in the graph, is not half the families.

Teacher note: This problem requires the student to view a graph and make inferences gained from comparing it with many other previously viewed and remembered graphs and their corresponding means and medians.

About the student work: From their work, one can see that students used the strategy of reasoning from the graph without actually modeling the data set with either a table or an organized list, that is, observed that the graph is skewed to the right and that therefore the mean is larger than the median.

Student Work

Student Response A

This response demonstrates the student's ability to read information from a graph and use it to answer the question posed. It is also indicates an understanding of how to determine both the mean and the median.

a) the median number of pets per family is more than the mean. not true
b) the mean number of pets per family is more than the median true
c) the mean number of pets per family is 2. not true
d) over half of the families had at least 6 pets. not true

1,1,1, 2, 2, 2, 2, 2, 2, 2, 2, 2, 2, 2, 2, 2, 2, 2, 2, 3, 3, 3, 3, 3, 3,
(3) 3, 3, 3, 4, 4, 4, 4, 4, 4, 4, 5, 5, 5, 6, 6, 6, 6, 6, 7, 7, 8, 9, 10,

all added 0, 12
= 189
median is 3
mean is 4 this is greater
189 ÷ 50 = 3.7 (4)   than this
round up

Student Response B

From the numbers on the graph and the conclusion, the student appears to have reasoned from the total number of pets and not the number of pets per family. In this instance, the faulty reasoning did not hinder the results. The student got the right answer for the wrong reason.

a) the median number of pets per family is more than the mean.
b) the mean number of pets per family is more than the median
c) the mean number of pets per family is 2.
d) over half of the families had at least 6 pets.

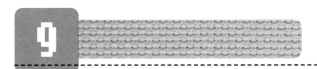

there are a lot of high numbers so that the mean will be higher than the median.

9

One class made the graph below to show the months in which they were born. Jennifer said that the mode of the classmates' birth months is 7. Is Jennifer correct?

Why or why not?

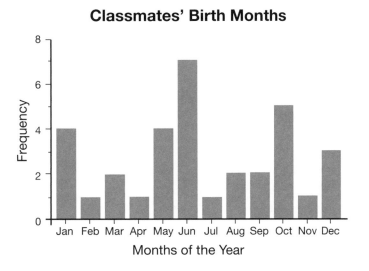

Classmates' Birth Months

About the mathematics: This item involves the ability to distinguish the data value or categorical data value from the frequency when defining the mode.

Solution: The correct response is June. The students who answered 7 for the mode were using the frequency rather than the numerical, or categorical, data. The value of 7 helps one select the category, but the category itself defines the mode. When analyzing categorical data, the category having the maximum frequency is called the *model category*.

Rubric (1 possible point)

1 point: Gives correct response and complete explanation

0 points: Gives incorrect response or no response

About the student work: To determine the mode, many students in the pilot test looked for the number that appears most often. This method works for a set of numerical data but not when the information is in graphical form.

Student Work

Student Response A

This response shows a clear understanding that the mode is categorical.

> No, Jennifer isn't correct if she needed to find the mode she would have to say June not 7. What Jennifer would use 7 for is using it as the frequency so June is the mode & 7 is the frequency.

Student Response B

This response indicates an understanding of how to determine the mode from the frequency of the bars on the graph; for example, four of the bars represent a month containing only one birthday.

> No because 1 is the mode. 7 only apears once and 1 apears the most (4 times).

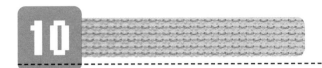

The track coach has recorded the times of the top four runners in the 100-meter race this year. The following table gives the times for each girl. Only two girls may compete in the district track meet. Which two girls would you select for the meet, and why? (Reminder: the lower times are preferred.)

Race Times in Seconds

Runner	Race 1	Race 2	Race 3	Race 4	Race 5	Race 6	Race 7
Ashley	15.9	15.0	14.2	15.3	14.5	14.8	14.7
Dara	15.6	15.5	14.8	15.1	14.5	14.7	14.5
Tanisha	15.8	15.7	15.4	15.0	14.8	14.6	14.5
Suzie	15.2	14.8	15.0	14.7	14.3	14.9	14.5

About the mathematics: This item involves the abilities to reason from a table of values, to compute the mean and median, and to use them in an argument.

Solution: Solutions will vary.

Runner	Minimum	Lower Quartile	Median	Upper Quartile	Maximum	Mean
Ashley	14.2	14.6	15	15.45	15.9	14.9
Dara	14.5	14.5	14.5	15.05	15.6	14.9
Suzie	14.3	14.5	14.7	14.95	15.2	14.7
Tanisha	14.5	14.75	15	15.4	15.8	15.02

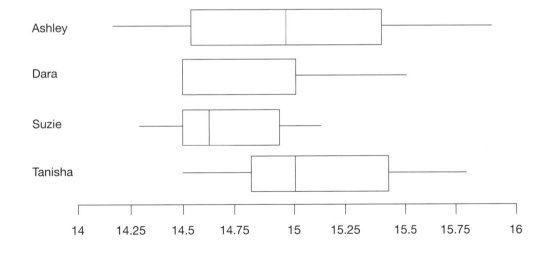

Some possible explanations follow:

a. One could choose Suzie and Tanisha because Suzie had the lowest mean time for the seven races and has five of the seven times below 15 seconds, whereas Tanisha shows continuous improvement, ending with her last time equaling that of Suzie and Dara.

b. One could choose Suzie and Dara because Suzie had the lowest mean time for the seven races and has five of the seven times below 15 seconds, whereas Dara had a mean time just two seconds higher than Suzie's and her median is lower than that of either of the other girls.

c. One could choose Suzie and Ashley because Suzie had the lowest mean time for the seven races and has five of the seven times below 15 seconds, whereas Ashley had the fastest time of any of the girls and is showing improvement.

Rubric (2 possible points)

2 points: Makes good use of numerical statistics and presents a reasonable argument

1 point: Gives a reasonable explanation for the choice but no numerical argument

0 points: Makes no response or gives a nonsensical explanation

About the student work: Their work suggests the students in the pilot test did not apply statistical representation or arguments in a meaningful way.

Student Work

Student Response A

In this response, the reasoning from the numbers is intuitive. This process can be beneficial in the development of conjecturing.

Race Times in Seconds

RACE #	1	2	3	4	5	6	7	
SUZIE	15.2	14.8	15.0	14.7	14.3	14.9	14:5	5
TANISHA	15.8	15.7	15.4	15.0	14.8	14.6	14.5	
DARA	15.6	15.5	14.8	15.1	14.5	14.7	14.5	u
ASHLEY	15.9	15.0	14.2	15.3	14.5	14.8	14.7	u

I would want Suzie & Ashlly because the have lower numbers then other girls so that means they got done faster

Student Response B

The argument in this response uses the mean or average. The strikeout suggests that the student's first thought was that the higher times were the better times.

Race Times in Seconds

RACE #	1	2	3	4	5	6	7	
SUZIE	15.2	14.8	15.0	14.7	14.3	14.9	14.5	≈14.8
TANISHA	15.8	15.7	15.4	15.0	14.8	14.6	14.5	≈15.1
DARA	15.6	15.5	14.8	15.1	14.5	14.7	14.5	≈15
ASHLEY	15.9	15.0	14.2	15.3	14.5	14.8	14.7	≈14.9

I'd pick Suzie & Ashley because they have the two fastest average times

Standard: Select the appropriate statistical methods to analyze data

Expectation: Discuss and understand the correspondence between data sets and their graphical representations, especially histograms, stem-and-leaf plots, box plots, and scatterplots.

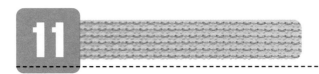

According to *The World Almanac 2001,* page 971, the American League strikeout leaders for each year from 1990 to 2000 were as shown in the following table:

Year	Player	Team	Strikeouts
1990	Nolan Ryan	Texas	232
1991	Roger Clemens	Boston	235
1992	Randy Johnson	Seattle	241
1993	Randy Johnson	Seattle	308
1994	Randy Johnson	Seattle	204
1995	Randy Johnson	Seattle	294
1996	Roger Clemens	Boston	257
1997	Roger Clemens	Toronto	292
1998	Roger Clemens	Toronto	261
1999	Pedro Martinez	Boston	313
2000	Pedro Martinez	Boston	284

a. Make a scatterplot of the data, and identify any trends or departure from a trend that can be seen.

b. What additional questions that are raised by the data would you like to investigate?

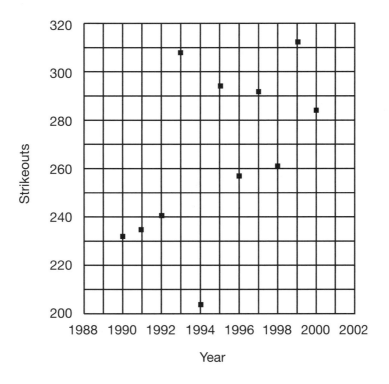

About the mathematics: This item involves making predictions or identifying trends from a graph of data.

Solution: Answers will vary.

a. The trend appears to be that the leading strikeouts per year have been increasing. The data points for 1993 and 1994 are of particular interest. Students might question what caused the number of strikeouts in 1993 to be so much higher and the strikeouts in 1994 to so much lower than the general trend.

b. Students might wish to investigate the strike-zone change, rule changes, the length of the season, or ball specifications. They should find, for example, that in 1994 a player's strike occured that shortened the season and cancelled the World Series.

Rubric (3 possible points)

3 points: Creates graph, identifies trend, and poses additional question(s)

2 points: Creates graph and identifies trend only

1 point: Creates graph only

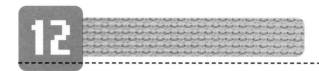

The following graph was posted in a local newspaper. No information was given with this graph. You have been asked to interpret and summarize the information contained within the scatterplot.

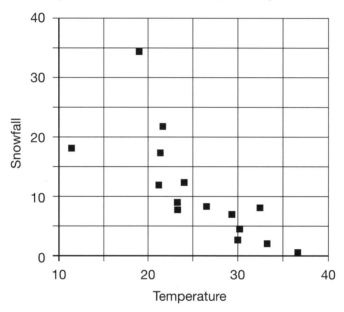

January Snowfall Columbus, Ohio (1974–1989)*

*Data from *Climatological Data, Ohio* monthly reports

Source: Adapted from *Data Analysis, an Introduction* by Jeffrey A Witmer (Englewood Cliffs, N.J.: Prentice Hall, 1992, pp. 48, 61)

About the mathematics: This item requires students to create and interpret scatterplots.

Solution: Answers will vary. The negative slope of the graph implies that little snowfall occurs in Columbus in warm Januarys. In one winter with an average temperature around 18° F, the amount of snow was much greater than typical. We can also see that the January snowfall total for Columbus ranged from just a couple of inches to around 20 inches, whereas the distribution of monthly average temperatures was usually 20° F to 30° F. The snowfalls greater than 15 inches spread out more than the lesser snowfalls.

About the student work: The results indicate a contentedness to identify one attribute.

Student Work

Student Response A

In this response, the student generalizes the slope contextually.

As the temperture goes up the amount of snow goes down

Student Response B

In this response, the student partitions the graph's information above and below 10 inches.

THERE ARE MORE snowfalls UNDER 10 inCHES

Student Response C

In this response, the student generalizes the slope contextually, identifies an outlier, partitions the information above and below 10 inches, and counts those below.

higher temperatures have less snow

There was one snowfall of about 34 inches

9 snowfalls were under 10 inches

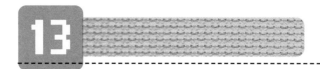

The tables list the age and height data for the 1995 rosters of two professional basketball teams, the Houston Rockets and the Chicago Bulls.

Houston Rockets

Player	Age	Height (cm)
Breaux	25	198
Brown	27	200
Cassell	26	188
Chilcutt	27	208
Drexler	33	198
Elie	32	193
Herrera	29	203
Horry	25	205
Jones	38	203
Maxwell	30	190
Murray	24	198
Olajuwon	32	210
Smith	30	188
Tabak	25	210

Chicago Bulls

Player	Age	Height (cm)
Armstrong	28	185
Blount	26	205
Buenchler	27	195
Harper	31	195
Jordan	33	198
Kerr	30	188
Krystakowiak	31	203
Kukoc	27	208
Longley	26	215
Myers	32	195
Perdue	30	210
Pippen	30	198
Simpkins	23	205
Wennington	32	210

Compare the distributions of the ages for the two teams and the heights for the two teams by using numerical summaries and plots of the data. Write a short statement explaining the differences and similarities.

Source: Adapted from *Connected Mathematics,* "Samples and Populations" (Upper Saddle River, N.J.: Pearson Prentice Hall, p. 56)

About the mathematics: This item involves reasoning statistically and creating a graphic representation based on the data presented.

Solution

Name	Minimum	Lower Quartile	Median	Upper Quartile	Maximum
Rockets Age	24	25	28	32	38
Bulls Age	23	27	30	31	33
Rockets Height	188	193	199	205	210
Bulls Height	185	195	200.5	208	215

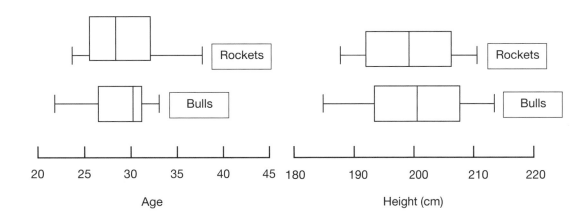

Comparing the box plots of the age data suggests that the Bulls players are a little older than the Rockets players on the average. By consulting the table of the values used in the box plots, one can see that the median age of the Bulls is in fact higher than that of the Rockets. The Bulls have the youngest player, and the Rockets have the oldest. The box plots indicate that the Bulls are closer in age than the Rockets.

The comparison of the box plots of the height shows that the Bulls players are slightly taller than the Rockets players. The Bulls' median height is just above 200 cm, which is slightly more than the Rockets' median height of 199, although the Bulls have the shortest and tallest player.

Rubric (2 possible points)

2 points: Makes good use of numerical statistics and presents a reasonable argument

1 point: Gives a reasonable explanation for the choice but no numerical argument

0 points: Makes no response or gives a nonsensical explanation

Standard: Develop and evaluate inferences and predictions that are based on data.

Expectation: Make conjectures about possible relationships between two characteristics of a sample on the basis of scatterplots of the data and approximate lines of fit

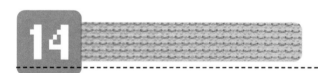

The scatterplot shows the length and mass of 10 pencils. Describe the slope of this graph in terms of the context.

Pencil Length versus Mass

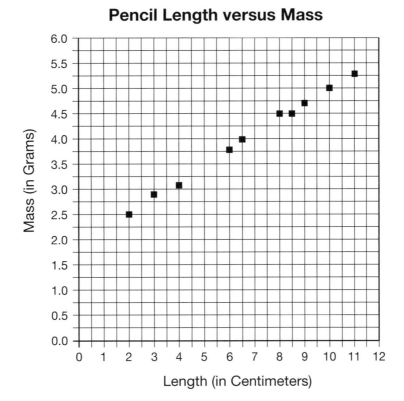

Source: Adapted from Massachusetts Mathematics Retest (July/November 2003 released items, problem 29, page. 53, problem 29)

About the mathematics: This item involves the ability to retrieve information from a graph and reason about the relationships so expressed.

Solution: Answers will vary. (Note: The student should realize that the slope is less than 1). As the length of the pencil increases 1 centimeter, the mass of the pencil increases about 0.3 grams.

About the student work: The solutions indicate students' difficulty in using the given scale and the inability to answer the question on the basis of the context.

Student Work

Student Response A

The work appears to indicate a lack of attention to the question asked, but it shows the student's memorization of slope as "over and up."

for every two you go over you go up two

Student Response B

The work seems to indicate the ability to determine the slope contextually, but the student did not read the scale properly.

FOR EVERY CENTIMETER INCREASE IN LENGTH IT GET 1 MORE GRAM OF MASS

Student Response C

This solution shows good understanding of the problem.

because the length increases by 1 centimeter the mass increases by $\frac{1}{2}$ gram

Standard: Develop and evaluate inferences and predictions that are based on data

Expectations: Make conjectures about possible relationships between two characteristics of a sample on the basis of scatterplots of the data and approximate lines of fit

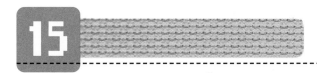

Using the information in the table below, create a scatterplot and line of best fit, then write conjectures about the relationship between the two characteristics of the chestnut oak trees grown in poor soil.

Age of Tree in Years	Diameter in Inches
4	0.8
5	0.8
8	1
8	2
8	3
10	2
10	3.5
12	4.9
13	3.5
14	2.5
16	4.5
18	4.6
20	5.5
22	5.8
23	4.7
25	6.5
28	6
29	4.5
30	6
30	7
33	8
34	6.5
35	7
38	5
38	7
40	7.5
42	7.5

Original data source: Chapman, Herman H., and Dwight B. Demeritt, *Elements of Forest Mensuration* (Albany, N.Y.: J. B. Lyon Co., 1932)

Source: Adapted from *Exploring Data,* (Palo Alto, Calif.: Dale Seymour Publications, © 1987 by The Bell Telephone Laboratories, application 34, p. 136, application 34)

About the mathematics: This item involves creating a scatterplot and line of best fit and conjecturing about any relationship between the variables.

Solution

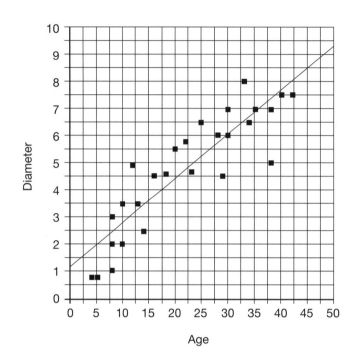

One would judge from the graph that the relationship appears not to be linear. More points fall above the line than below, and most of those that fall below are at either end of the line. This arrangement would lead one to think that the data have a curved shape.

Standard: Develop and evaluate inferences and predictions that are based on data

Expectation: Use conjectures to formulate new questions and plan new studies to answer them

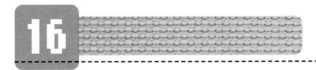

Suppose you wanted to survey the eighth grade class in your school to determine the relationship between watching television and grades.

a. What factors must we be concerned about in the design of our survey?
b. What questions would you ask?
c. What other factors might influence grades?

> **About the mathematics:** This item involves formulating questions and making conjectures. The following are representative correct responses.
> **Solution:** Solutions will vary.
> a. That the sample is truly a random sample. That the questions are not biased. That we survey an appropriate number of people.
> b. What is your current grade average, and approximately how many hours do you watch television during a week?
> c. The number hours you study during a week. The types of programs you watch. If you have a job, how many hours you work. The amount of sleep you get each night. Whether you eat breakfast.

Standard: Understand and apply basic concepts of probability

Expectation: Understand and use appropriate terminology to describe complementary and mutually exclusive events

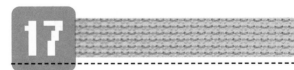

Suppose that A represents the following event:

No students were absent from study hall today.

Explain what the complement of A represents.

Source: Adapted from *Probability through Data* (Palo Alto, Calif.: Dale Seymour Publications, © 1999 by Addison Wesley Longman, p. 70)

About the mathematics: This item requires an understanding of complementary events in probability.

Solution: The complement is the event that one or more students were absent from study hall today.

About the student work: Students' work on this problem exemplifies a reasonable understanding of the concept.

Student Work

Student Response A

This response indicates no understanding of the concept.

EVERYBODY WHO IS THERE

Student Response B

This response leaves interpretation to the reader.

THE OPPOSITE

Standard: Understand and apply basic concepts of probability

Expectation: Use proportionality and a basic understanding of probability to make and test conjectures about the results of experiments and simulations

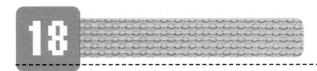

If you toss a fair coin eight times and get heads each time, which of the following is true?

a. The probability of the getting a head on the next toss is more that $1/2$.
b. The probability of the getting a tail on the next toss is more than $1/2$.
c. The probability of the getting a tail on the next toss is $1/2$.
d. The probability of the getting a head on the next toss is greater than the probability of getting a tail.

> **About the mathematics:** This item identifies students misconceptions about probability.
>
> **Solution:** The correct answer is c. The number of times a head or a tail appears in succession has no bearing on the fact that the probability on any one toss of getting a head or tail remains at $1/2$.

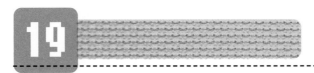

Suppose for sake of argument the fact is known that with no medication administered, half of those suffering from of a specific disease die and half recover. A doctor proposes to demonstrate that a new drug improves the recovery rate. She conducts a test on 200 patients. What conclusions can she make if 105 of the participants in the experiment recovered and 95 died?

> **About the mathematics:** This item involves making conjectures about an experiment on the basis of comparison with known probabilities or a simulation.
>
> **Solution:** When compared with the outcome of many trials of tossing a coin, 105 successes are not unusual, that is, it could have happened by chance; so the outcome of taking the new drug appears not to be any different from chance.

> **Standard:** Understand and apply basic concepts of probability

> **Expectation:** Compute probabilities from simple compound events, using such methods as organized lists, tree diagrams, and area models

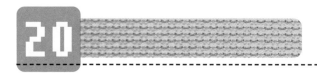

The table below summarizes by gender the number of musicians in an orchestra who wear glasses.

	Men	Women	Total
Glasses	32	3	35
No Glasses	56	39	95
Total	88	42	130

A member of the orchestra is chosen at random.

a. What is the probability that the person wears glasses? Explain your reasoning.
b. If you were told that the person is a man, would you change your answer to part a? Explain your reasoning.

> **Source:** Adapted from *Great Expectations,* Mathematics in Context (Austin, Tex.: Holt, Rinehart & Winston, © 1998 by Encyclopedia Britannica, p. 23)
> **About the mathematics:** This item requires understanding the relationships among relative frequencies, conditional probability, and dependent and independent events.
> **Solution**
> a. From the table, the probability that a person wears glasses is $35/130$ = 0.27, or about 27 percent.
> b. Yes. If the person is known to be a man, the probability is $32/88$ = 0.364, which is about 36 percent.

Rubric (2 possible points)

2 points: Involves the correct numbers and presents a valid explanation

1 point: Involves the correct numbers but gives no explanation or a confusing explanation

0 points: Attains incorrect solution or makes no attempt at a solution

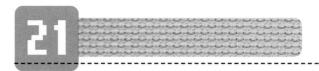

A government agency was interested in how the description of workers in the United States changed from 1985 to 1995. The table below shows the relative frequencies of new workers from 1985 to 1995. For example, 0.15 means that $^{15}/_{100}$ were white men.

Description of Workers

	White	Nonwhite	Immigrant	Total
Men	0.15	0.07	0.13	0.35
Women	0.42	0.14	0.09	0.65
Total	0.57	0.21	0.22	1.00

What is the probability that a randomly selected new worker is–

i.   an immigrant?
ii.   a woman?
iii. an immigrant and a woman?
iv.   an immigrant or a woman?
v.   white or nonwhite?
vi.   a man or nonwhite?
vii. nonwhite or an immigrant?

> **Source:** Adapted from *Probability through Data* (Palo Alto, Calif.: Dale Seymour Publications, © 1999 by Addison Wesley Longman, p. 57)
>
> **About the mathematics:** This item requires that students understand the relationship between relative frequencies and probability and know the addition rule.

Solution

   i. 0.22

   ii. 0.65

   iii. 0.09

   iv. $0.22 + 0.65 - 0.09 = 0.78$; the woman immigrant is counted twice.

   v. $0.57 + 0.21 = 0.78$.

   vi. $0.35 + 0.21 - 0.07 = 0.49$; the nonwhite man is counted twice.

   vii. $0.21 + 0.22 = 0.43$.

Rubric (2 possible points)

   2 points: Involves the correct numbers and presents a valid explanation

   1 point: Involves the correct numbers but gives no explanation or a confusing explanation

   0 points: Arrives at incorrect solution or makes no attempt at a solution

About the student work: None of the work indicates an awareness of the addition rule or union of sets. The work shows no evidence of understanding that in the "or" situation, some individuals are counted twice.

Student Work

Student Response A

This response indicates a good understanding of the table values; however, how the student arrived at the 144% or 44% in the first entry is unclear. Parts iv and vi deal with the addition rule, which the student apparently has not learned.

   i.   an immigrant? 144%

   ii.   a woman? 65%

   iii.   an immigrant and a woman? 9%

   iv.   an immigrant or a woman? 87%

   v.   white or nonwhite? 78%

   vi.   a man or nonwhite 56%

   vii.   nonwhite or an immigrant? 43%

Student Response B

The response to part ii is the sum of all values listed for women in the table. Some answers indicate understanding, whereas others reveal a lack of understanding.

    i.   an immigrant? .22

    ii.   a woman? 1.3

    iii.   an immigrant and a woman? .09

    iv.   an immigrant or a woman? 1.3, .09

    v.   white or nonwhite? .78

    vi.   a man or nonwhite .7

    vii.   nonwhite or an immigrant? .21, .22 .43

Student Response C

This response indicates confusion when items are combined.

    i.   an immigrant? 22%

    ii.   a woman? 65%

    iii.   an immigrant and a woman? 87%

    iv.   an immigrant or a woman? 22% or 65%

    v.   white or nonwhite? 57% or 21%

    vi.   a man or nonwhite 35% or 21%

    vii.   nonwhite or an immigrant? 21% or 22%

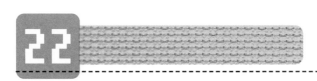

A basketball player shoots foul shots with a two-thirds accuracy record. That is, he has scored a basket in two out of every three attempts. He is given a free throw from the foul line and is given a second shot only if he scores a basket on the first shot. In this one-and-one situation, he can score 0, 1, or 2 points. Determine the approximate probability that he will score 2 points. Show your work.

Source: Adapted from *The Art and Techniques of Simulation* (Palo Alto, Calif.: Dale Seymour Publications, 1987, p. 44, application 22)

About the mathematics: This item involves understanding the probability of compound events.

Solution: $^2/_3 \times {}^2/_3 = {}^4/_9$. Two other forms of the solution are as follows:

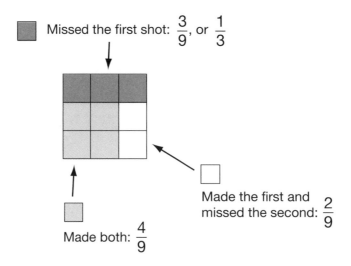

Missed the first shot: $\dfrac{3}{9}$, or $\dfrac{1}{3}$

Made the first and missed the second: $\dfrac{2}{9}$

Made both: $\dfrac{4}{9}$

Or:

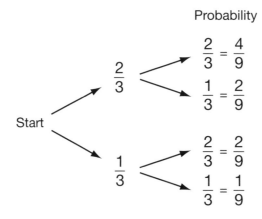

Probability

Start

$\dfrac{2}{3}$ → $\dfrac{2}{3} = \dfrac{4}{9}$

$\dfrac{1}{3} = \dfrac{2}{9}$

$\dfrac{1}{3}$ → $\dfrac{2}{3} = \dfrac{2}{9}$

$\dfrac{1}{3} = \dfrac{1}{9}$

Professional Development

THE PURPOSE of this chapter is to offer guidance to in-service and preservice teachers in the areas of understanding and using levels of complexity, using and adapting multiple-choice items, using assessment tasks as an in-service topic, using scoring rubrics, using technology in assessment, and designing assessment items.

Levels of Complexity

How difficult a problem is for a student to solve is a function of how familiar the student is with the mathematics being assessed, and what the student is asked to do with the mathematics. Thinking about this interaction is essential for both instruction and assessment.

> **What does a student mean when they say, "This is a really hard problem"?**

The 2005 Mathematics Framework for the National Assessment of Educational Progress (NAEP) addressed this issue by defining levels of complexity. The NAEP Levels of Complexity address the *"demands on thinking"* that an item makes, assuming that the student is familiar with the mathematics being assessed. The Levels of Complexity do not try to capture the different approaches that students use, particularly approaches taken when the mathematics is unfamiliar.

The NAEP 2005 Framework defines three Levels of Complexity: low, moderate, and high.

General Descriptions of NAEP Levels of Complexity[1]

Low Complexity

This category relies heavily on "recall and recognition of previously learned concepts and principles." The following are some, but not all, of the demands made by items in the low-complexity category as described in the NAEP 2005 Framework:

- Recall or recognize a fact, term, or property
- Recognize an example of a concept
- Compute a sum, difference, product, or quotient
- Recognize an equivalent representation
- Perform a specified procedure
- Evaluate an expression in an equation or formula for a given variable
- Solve a one-step word problems
- Draw or measure simple geometric figures
- Retrieve information from a table or graph

Moderate Complexity

Items in the moderate-complexity category "involve more flexibility of thinking and choice alternatives than do those in the low-complexity category. They require a response that goes beyond the habitual and ordinarily has more than one step." The following are some, but not all, of the demands made by items in the moderate-complexity category as described in the NAEP 2005 Framework:

- Represent a situation mathematically in more than one way
- Select and use different representations, depending on the situation
- Solve a word problem requiring multiple steps
- Compare figures or statements
- Provide justifications for steps in a solution process
- Interpret a visual representation
- Extend a pattern
- Retrieve information from a graph, table, or figure and use it to solve a problem requiring multiple steps
- Formulate a routing problem, given data and conditions
- Interpret a simple argument

High Complexity

High-complexity items "make high demands on students, who must engage in more abstract reasoning, planning, analysis, judgment, and creative thought. "The following are some, but not all, the demands made items in the high-complexity category as described in the NAEP 2005 Framework:

---

1 National Assessment Governing Board, *Mathematics Framework for 2005* (Prepublication Edition, December 2001)

- Describe how different representations can be used for different purposes
- Perform a procedures having multiple steps and multiple decision points
- Analyze similarities and differences between procedures and concepts
- Generalize a pattern
- Formulate an original problem, given a situation
- Solve a novel problem
- Describe, compare, and contrast solution methods
- Formulate a mathematical model for a complex situation
- Analyze the assumptions made in a mathematical model
- Analyze a deductive argument
- Provide a mathematical justification

Although the foregoing levels were designed for use in the development of assessment items for NAEP, they have direct implications for classroom assessment and instruction. The assumption underlying these levels is that what a student does with the mathematics that he or she learns is important. If students are asked only to recall information or to perform routine procedures instructionally, they will not likely be able to solve complex problems related to the same mathematics.

In addition, the three Levels of Complexity do not include a "fourth" level that also has direct implications for classroom instruction and assessment. This level, described by Webb[2], involves "complex reasoning, planning, and thinking over extended periods of time." Such instructional and assessment activities as projects and statistical studies fall into this important category.

Here are some ways that teachers can use the Levels of Complexity:

- Use the examples throughout this sampler to help understand the levels.
- During instruction, include questions that span the range of complexity, including Webb's fourth level.
- Use the levels to develop classroom tests and quizzes that include a range of complexity.
- More information about the NAEP Levels of Complexity and Webb's Depth of Knowledge Levels can be found in NAEP (2001) and Webb (1997, 2002).

Using and Adapting Multiple-Choice Items

Frequently students' mathematical ability is judged by their responses to multiple-choice items. Because multiple-choice items are easy to score, they can provide useful information about large groups of students and their progress toward particular standards (Concept to Classroom 2004; Shannon 1996). Some multiple-choice items probe deeply into the conceptual knowledge that a student brings to the task. In general, however, multiple-choice items provide little insight into students' actual abilities and are not capable of assessing the full range of student activities

---

2 Webb, Norman, *Criteria for Alignment of Expectations and Assessments on Mathematics and Science Education,* Research Monograph Number 6 (Washington, D.C.: Council of Chief State School Officers, 1997).

described in *Principles and Standards for School Mathematics* (NCTM 2000; Shannon 1999; Stenmark 1996; National Research Council 1993). Students who will be judged using multiple-choice items should have the opportunity to experience such items; however, for these item to be useful in terms of understanding individual student approaches and to influence teaching, they should be used in ways that will offer an instructional advantage.

One problematic aspect of a multiple-choice item is the possibility that a student will get the correct answer either by guessing or through incorrect reasoning. For example, consider the following problem from the Virginia Standards of Learning Assessment (VDE 2002):

> **If $0.3 < x < 35\%$, which of the following could be the value of x?**
> A. $1/4$
> B. $1/3$
> C. $1/2$
> D. 1

A student might choose the correct solution, B, simply because all the numbers have a 3 in them. Similarly, a student in the pilot test, when asked to explain the reason for picking a particular solution, gave the following response:

> I don't know, they all look the same and they all have the same numbers and I normally pick the second to last.

From another perspective, correct responses to multiple-choice items do not help the teacher gauge the level at which a student understands the topic. Students may be on different levels in their approaches, use of strategies, and schematizations (ven den Heuvel-Panhuizen 1996) that are not revealed by a single-response item. For example, consider the following item from the 1990 NAEP assessment:

> Which of the following is true about 87 percent of 10?
> A. It is greater than 10. D. Can't tell.
> B. It is less than 10. E. I don't know.
> C. It is equal to 10.

The actual sequence of skills that students use to solve the problem is impossible to determine. One student may consider the item procedurally, multiplying 10 by 0.87, whereas another may reflect that 87 percent of a quantity is less than the quantity itself. Without more information, how a student has arrived at the solution is impossible to know, even when the reasoning is correct.

Clearly, multiple-choice items allow no window into the way students are thinking and without follow-up will provide little information that can be used to guide instruction. In a classroom setting this shortcoming can be addressed by asking students to explain their choices as part of the task itself. The multiple-choice tasks in this sampler are accompanied by students' explanations collected from the students in the pilot test, who were instructed to explain how they made their choices. Presenting multiple-choice tasks in this way supports instruction by giving teachers insights into students' thinking about both incorrect and correct solutions.

Multiple-choice items can also lend a start to writing more open-ended items. Many multiple-choice items can be rewritten to be open-ended simply by eliminating the choices. The problem in figure 6.1 is presented in both formats. Notice, however, that using the open format may not yield any information about student thinking; to do so, the item requires that students explain how they arrived at their conclusion and to justify their thinking. Changing the question to "Explain how you would find the average snowfall (in inches) for Cleveland during that time period" will broaden the focus of the task to include both the answer and a description of how students approach the problem.

| The seasonal snowfalls in Cleveland, Ohio, for the winters 1970-71 through 1980-81 to the nearest inch were as follows:

51, 46, 69, 59, 67, 54, 63, 90, 38, 39, 61.

What is the average snowfall (in inches) for Cleveland during that time period? (Round your answer to nearest inch.)

    a) 58
    b) 54
    c) 57
    d) 64 | The seasonal snowfalls in Cleveland, Ohio, for the winters 1970-71 through 1980-81 to the nearest inch were as follows:

51, 46, 69, 59, 67, 54, 63, 90, 38, 39, 61.

What is the average snowfall (in inches) for Cleveland during that time period? (Round your answer to nearest inch.) |

Data from *Climatological Data, Ohio* monthly reports

Fig. 6.1. A problem presented in both multiple-choice format and open-ended format

Other problems might need more adjustment. Consider the following eighth-grade item from The Massachusetts Comprehensive Assessment System (MDE 2001, p. 299.

Use the balance scale below to answer question 15.

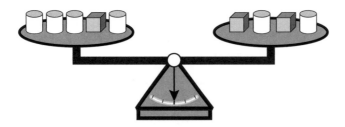

15 Which of the following shows the relationship between the weights of one cylinder and one cube?

✓ A. One cube weighs the same as two cylinders.
   B. One cube weighs the same as four cylinders.
   C. One cylinder weighs the same as two cubes.
   D. One cylinder weighs the same as four cubes.

In this instance, the question is appropriate for a more open response. Possible questions might include these:

- How can you find what weighs the same as one cube?
- Draw another, balanced scale and explain why it is balanced.
- Describe the relationship between the weight of a cylinder and the weight of a cube?

Each of these questions addresses the same basic content as the original item but asks students to explain the thinking they bring to the problem.

Stenmark (1996) suggests that problems can be made open ended by changing the question as illustrated previously, either by introducing additional information or extra questions or by deleting some necessary information. She advocates looking for ways to "see student thinking rather than test-writer thinking" (p. 20) and provides a list of question ideas to help achieve this outcome, including "How would you explain this process to a younger child?" "How were you sure your answer is right?" and "Explain in writing or with a diagram what this means."

The box plot represents the test scores of a middle school mathematics class.

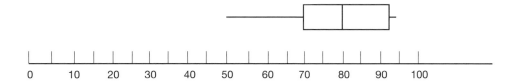

Which of the statements could be true of grades earned by the class on the test?

a) Three fourths of the class had a grade lower than 80.
b) Half of the class had a grade between 90 and 70.
c) Three fourths of the class had grades 70 or above.
d) Half of the class had a grade between 70 and 80.

Source: *Mathematics Assessment Sampler, Grades 6–8: Items Aligned with NCTM's "Principles and Standards for School Mathematics"* (Reston, Va.: National Council of Teachers of Mathematics, 2005, p. 192)

This item can be opened to foster better understanding of student thinking by including the request "explain why you made your choice" after each choice.

Multiple-choice tests cannot provide all the necessary information on students' mathematical attainments, progress, and capabilities. Appropriately used and adapted, however, they can give students opportunities to perform, allow teachers to make inferences about what students know and can do in mathematics, and contribute to the broader goal of improving mathematics teaching and learning.

Using Assessment Tasks

Professional Standards for Teaching Mathematics (NCTM 1991) suggests that mathematics and mathematics education instructors in preservice and continuing education programs should model good mathematics teaching by–

- posing worthwhile mathematical tasks;
- engaging teachers in mathematical discourse;
- enhancing mathematical discourse through the use of a variety of tools, including calculators, computers, and physical and pictorial models;
- creating learning environments that support and encourage mathematical reasoning and teachers' dispositions and abilities to do mathematics;
- expecting and encouraging teachers to take intellectual risks in doing mathematics and to work independently and collaboratively;
- representing mathematics as an ongoing human activity;
- affirming and supporting full participation and continued study of mathematics by all students. (p. 127)

Professional development programs should engage teachers in the process of "doing" mathematics in a manner similar to that which they will use with their students. Mathematics teachers' experiences have a direct impact on the type of mathematical experiences they will provide for their students. Teachers who have limited experience in an inquiry-based learning environment are not likely to provide such an environment for their students. A rich mathematical task presents an opportunity to engage teachers in a discussion about ways that they reasoned about the mathematics in the task. By engaging in discussions with their peers, teachers are afforded an opportunity to encounter different ways of reasoning about the mathematics in a problem. Teachers leave the professional development experience feeling more confident in their own abilities to do mathematics and more willing to explore using rich tasks with their own students.

Using High-Quality Tasks

The following approaches are helpful when using high-quality tasks for purposes of professional development:

Teachers work the task using the mathematical knowledge and skills that they think students are likely to bring to a task. Teachers are encouraged to work the task individually at first. This approach gives teachers time to choose their own solution strategy and not be influenced by others' thinking. While teachers are working on the task, the facilitator moves around the room looking for a variety of ways in which teachers reasoned about the mathematics. At this time the facilitator can attend to teachers who may be having some difficulty with the task. Teachers may find this approach less threatening than a large-group dynamic and be more likely to engage with the task.

Teachers discuss the different ways that they reasoned about the situation and the mathematics. After teachers have had sufficient time to work independently, the facilitator brings the group together to share the ways that they reasoned about the mathematics. The facilitator listens carefully to each presentation and asks participants to consider the reasonableness of each solution. Some solutions may raise interesting issues about teaching and learning. Depending on time constraints and other factors, the facilitator may choose to ask teachers to discuss the instructional implications of the various approaches. Some teachers may present solutions that are beyond the mathematics that students might use in the task. In this situation, the facilitator uses the discussion to deepen teachers' content knowledge and help other teachers make connections. Teachers are encouraged to ask questions and challenge ideas that are presented.

Teachers consider what students in the middle grades might do with the task. While working on a task, teachers often consider what their students are likely to do with the task. This discussion helps teachers decide the appropriateness of the task, where it fits in their school's curriculum, and what mathematical experiences students need if they are to be successful on high-quality tasks.

Teachers examine student work to compare students' reasoning about the situations and approaches to solving the problems. The student work is selected so that it raises important issues about teaching and learning.

Teachers try tasks with their own students and bring sample work to subsequent professional development sessions.

Using Multiple-Choice Items

Teachers and districts are facing increased pressures and greater accountability requirements for student achievement since the No Child Left Behind (NCLB) Act of 2001 was enacted. Under NCLB, each state is required to measure students' mathematics progress in grades 3 through 8 and at least once during grades 10 through 12. Many states and districts use tests that include predominately multiple-choice items because such tests are economical to administer, perceived to be more reliable than other assessment instruments, and easier to score than other types of assessments. However, multiple-choice items may not reveal students' understandings and misconceptions. Without additional information, a student's reasoning about an item may be impossible to know, and the test serves little purpose for improving instruction.

The following approaches are useful when using multiple-choice items for professional development purposes to enhance learning:

Teachers work the item(s) using the mathematics they think students will bring to the items. Multiple-choice items typically do not require a great amount of time for teachers to complete. A more productive and stimulating approach may be to select a collection of items that focus on a specific mathematical process or strand. Another strategy is to remove the distracters and ask teachers to solve each item. This approach provides an opportunity to discuss the ways teachers reasoned about the mathematics. The teachers generate potential distractors and compare them with the distractors in the item.

Teachers analyze and discuss the distractors to determine what each might reveal about a student's understandings or misconceptions. Teachers consider whether the distractors are reasonable or whether different distractors are more suitable.

Teachers examine students' written work to gain insight into students' thinking and reasoning. This analysis can be made only if students explain or show how they solve a problem. Students may arrive at a correct solution using incorrect or faulty reasoning.

Teachers consider how to use the items with students. Teachers offer suggestions for ways to engage students in a discussion about the mathematics. One approach is to ask students to consider the possible lines of reasoning behind a person's choice of each of the distractors.

Scoring Rubrics

Assessment in mathematics serves many purposes. One of the most important is to glean information to make appropriate instructional decisions. Another is to provide a progress report to students and their families. For either of these purposes, we need to develop a form of communication that allows both the teacher and the student, as well as other interested parties, to correctly interpret the information that is presented. Collecting and presenting assessment data have long traditions in all teaching, and mathematics teachers have typically given homework, quizzes, and tests to measure and report on students' progress.

Recently, teachers have developed scoring rubrics to foster consistency when marking student work. The simplest scoring is done using a holistic rubric. A holistic rubric describes the criterion for awarding a certain score to each type of response. Rubrics that use an all-or-nothing approach usually provide minimal information about students' abilities. More informative rubrics award points for steps along the way. For example, a 5-point rubric could award 0 points if nothing is written, 2 points if an attempt is made with little success, 3 points if a correct start is made, 4 points if a minor mistake is made, and 5 points for an exemplary response. This kind of partial-credit rubric often encourages a student to provide detailed information about the solution process, and it helps the teacher evaluate a student's progress, even when the answer is not completely correct. An excellent discussion of tasks and the use of rubrics is found in Stenmark (1991).

More detailed rubrics can focus on specific aspects of the problem-solving process. For example, teachers can give credit, or perhaps partial credit, on the basis of their observations of specific components in the student's work. Such rubrics may focus on communication of ideas, understanding the problem, using mathematics correctly, interpreting a solution, using appropriate technology, or many other possibilities. The establishment of specific criteria for scores that can be earned is essential to the effectiveness of scoring rubrics. For example, a student's understanding a problem can range from no indicated understanding, on one end of the scale, to a complete and verified solution on the other. In between, the student might indicate that he or she understood the problem but made little progress toward a solution. Points on this scale would reward students who explain their understanding of the problem, even if the solution is incomplete.

Analytic scoring rubrics include process dimensions to assess student work: Problem Solving, Communication, Representations, Connections, and Presentation. Teachers have reported vast improvement in students' ability when they assess specific skills through a focus on these dimensions. Students will make an effort to improve communication, both oral and written, when they know that communicating their thoughts and strategies is being assessed. This focus leads to two improvements, as noted previously: (1) students become more aware of important criteria, and (2) teachers attain better feedback on which to base their instructional decisions. For examples of multiple-dimension rubrics, see Bush and Leinwand (2000).

Technology and Assessment

Technology is essential in teaching and learning mathematics; it influences the mathematics that is taught and enhances students' learning.... Electronic technologies—calculators and computers—are essential tools for teaching, learning, and doing mathematics. They furnish visual images of mathematical ideas, they facilitate organizing and analyzing data, and they compute efficiently and accurately. They can support investigation by students in every area of mathematics, including geometry, statistics, algebra, measurement, and number. When technological tools are available, students can focus on decision making, reflection, reasoning, and problem solving. (NCTM 2000, p. 24)

The technology principle in the National Council of Teachers of Mathematics' (NCTM) *Principles and Standards for School Mathematics* (2000) offers a clear rationale for the use of technology in teaching and learning mathematics and reiterates the message of *Curriculum and Evaluation Standards for School Mathematics* (NCTM 1989) about the importance of the technology. In general, mathematics educators have supported the use of calculators in mathematics classrooms since the 1970s, but despite the potential offered by technology as a tool for learning mathematics, its use is still challenged, raising issues central to mathematics instruction (Fey 1989;

Usiskin 1999a). Does the use of technology help or hinder learning? What degree of understanding can be attained through repeated application of algorithms? What new understandings, if any, can arise from calculator use, and what understandings, if any, may be lost? How does the use of calculators interact with the development of basic skills? What content becomes obsolete because of the existence of calculators? What is meant by "appropriate" use of calculators? Over the past quarter of a century, little consensus has emerged on the answers to these questions.

Research results, however, have begun to foster some insights. Early meta-analysis studies (Hembree and Dessart 1986, 1992) found that at almost every grade level, the use of calculators enhanced basic-skill acquisition, improved problem solving, and resulted in better attitudes toward mathematics. Subsequent work supported these findings and indicated that calculator use had a positive effect on students' conceptual knowledge and their mathematical computation, and did not hinder their development of pencil-and-paper skills (Smith 1997). On the 2000 National Assessment of Educational Progress (NAEP), eighth graders whose teachers reported that they permitted unrestricted use of calculators had higher average scores than did students whose teachers restricted calculator use (Braswell 2001).

A synthesis of the research on handheld graphing calculator use reported that, depending on how the calculators were used, on what tasks, and the degree to which teachers integrated the tool with the mathematics, their use could make a difference in student achievement. In general, "[t]he research indicates that students who use handheld graphing technology with curriculum materials supporting its use had a better understanding of functions, variables, solving algebra problems in applied contexts, and interpreting graphs than those who did not use the technology. Students who spent more time using handheld graphing technology showed greater gains than students who had access to the technology for brief interventions or short periods of time" (Burrill et al. 2003, p. i).

Research on learning supports these findings, suggesting that technology-based tools can enhance student performance when they are integrated into the curriculum and used to support learning (National Research Council 1999). Thus, given that calculators can make a difference in student learning, integrating the calculator with the content in meaningful ways and exploiting the technology to enhance learning and to expand the mathematics that is possible to teach implies that calculators have the potential to shape what is taught.

Another central force in defining what is taught is what is assessed (Leinwand 1992; AERA 2000; Hess 2003), and past experience indicates that the link between technology and its use on tests has a major impact on the role of technology in the classroom. "To a significant extent, however unfortunate, the relationship between technology and examinations has a strong tendency to dictate the relationship between technology and mathematics generally throughout a course of study"

(Kissane 2000, p. 1). Shortly after the advent of graphing calculators in 1985 and in keeping with the vision of mathematics teaching and learning described in NCTM's *Curriculum and Evaluation Standards for School Mathematics* (1989), mathematics educators argued for changes in mathematics assessment, protesting against national, standardized instruments that test only shallow curricula (Lipson, Faletti, and Martinez 1990). *Assessment Standards for School Mathematics* (NCTM 1995) postulated that assessment should provide all students with the opportunities to formulate problems, reason mathematically, make connections among mathematical ideas, and communicate about mathematics. Clements (1998) advocated for additional forms of assessment, including those called "authentic," "performance," or "direct," based on tasks that require reflection, emphasize important concepts, and demand persistence. As students engage in such tasks with the use of calculators, the teacher can observe their work and focus on their thinking.

> Thus, technology aids in assessment, allowing teachers to examine the processes used by students in their mathematical investigations as well as the results, thus enriching the information available for teachers to use in making instructional decisions. (NCTM 2000, p. 26).

What content is appropriately tested with a calculator?

Shortly after graphing calculators appeared in school classrooms, a joint symposium of the College Board and the Mathematical Association of America recommended that mathematics achievement tests should be curriculum based and that no questions should be used that measure *only* calculator skills or techniques (Kenelly 1989). As calculators were integrated into assessments, the objectives tested fell into two categories: (1) calculator-specific, in which using the calculator is necessary to answer the question, and (2) calculator-neutral, in which the problem can be done with or without a calculator. Some educators argued for unrestricted use of calculators (Burrill 1999), whereas others were more cautious (Usiskin 1999b). The mathematics objectives commonly considered as appropriate for testing with calculators included—

a) the exploration of number patterns;
b) the use of a guess-and-check strategy for problem solving (Wilson and Kilpatrick 1989);
c) the process of hypothesis formation and verification (Heid 1988); and
d) the solution of problems using realistic data.

Calculators have typically been prohibited when assessing basic computational skills (Oregon State Department of Education 1985; Carter and Leinwand 1987; North Carolina State Department of Public Instruction 1983; Hopkins 1992), although even the Long Term Trend Study of computational skills carried out by NAEP since 1973 does allow calculator use on a few of the items (NCES 2000).

Items designed for testing can be greatly influenced by the graphing calculator. Figure 6.2 provides a typology that may be useful (adapted from Kemp, Kissane, and Bradley [1996, p. 3]).

Graphing calculators are expected to be used

1   Students are explicitly advised or even told to use graphing calculators.
2   Alternatives to the use of graphing calculators are inefficient.
3   Graphing calculators are used as scientific calculators only.

Graphing calculators are expected to be used by some students and not by others

4   Use and nonuse of graphing calculators are both suitable.

Graphing calculators are not expected to be used

5   Exact answers are required.
6   Symbolic answers are required.
7   Written explanations of reasoning are required (although work on the calculator may precede the written explanation).
8   The task involves extracting the mathematics from a situation or representing a situation mathematically.
9   The use of a graphing calculator is inefficient.
10   The task requires that a representation of a graphing calculator screen will be interpreted (work may contain a sketch of the screen).

Fig. 6.2. Expected use of graphing calculators on examinations

The types of distractors and the amount of time required to complete a task also have an effect on designing items for calculator-based tests. Should one include most common calculator errors as distractors? In studying the use of calculators in testing, Long, Reys, and Osterlind (1989) found that the selection of distractors representing the most common calculator errors "often produces responses that challenge the definition of plausible" (p. 323). Students using calculators on the NAEP seemed to spend more time on each test item than students not allowed to use calculators (Hopkins 1992). The studies do not clarify whether the extra time was needed because students might have been unfamiliar with the calculator or because of the types of items given on the test.

Should calculator use be optional or required during testing?

Advocates for the use of calculators on tests offer a variety of arguments. The use of calculators allows greater flexibility in the construction of test items. Using the calculator allows students to concentrate on strategies for solution and does not require them to rely on memorized computational algorithms. Access to a calculator removes computational barriers and gives more students the opportunity to demonstrate their conceptual understanding and ability to solve problems. Well-designed calculator investigations may reveal interesting or subtle mathematical concepts (NRC 2001). The use of calculators also allows for the use of realistic numbers, thereby improving the validity of items and allowing problem-solving situations to be more akin to those found in practice (Wilson and Kilpatrick, 1989). The joint 1986 Symposium on the Use of Calculators in Standardized Testing of the College Board and the Mathematical Association of America acknowledged that deciding whether to use a calculator when addressing any particular test item is itself an important skill. Consequently, the symposium findings recommended that calculators should not be required on all test items on a calculator-based mathematics test (Hopkins 1992).

Returning to the link between what is tested and what takes place in classrooms, an important point to note is that the use of technology, in particular, calculators, is significant. A 2000 national survey of mathematics and science teachers reports that nearly half the teachers have their students use calculators or computers to develop conceptual understanding at least once a week (Hudson et al. 2002). Approximately half of middle school teachers and three-fourths of high school teachers report that calculators are fully integrated into their curriculum. The 2000 NAEP reports that more than 85 percent of eighth-grade students and 90 percent of twelfth-grade students had access to calculators in their mathematics classes (NCES 2000). The College Board of Mathematical Sciences (Lutzer et al. 2001) survey of two- and four-year colleges shows a dramatic increase in the use of graphing calculators since 1995.

What is the status of calculators and assessment? According to the NAEP data, more than 65 percent of eighth-grade teachers reported that they permitted calculator use on class assessments. A noteworthy finding of the 2000 NAEP, which permits calculators on certain sections of the assessment, was that eighth-grade students of those teachers had higher average NAEP scores than students whose teachers did not permit calculator use on tests (NCES, 2001). In 2004, 26 states permitted or required graphing technology on state assessments (Tolbert 2004). Calculators are allowed or required on the ACT, the National Assessment of Educational Progress (NAEP), Advanced Placement examinations, and SAT II: Math subject tests.

In conclusion, the NRC Committee on the Developments in the Science of Learning reports that technology makes possible a motivating curriculum based on real-world problems, provides scaffolds and tools to enhance learning, and gives students and teachers opportunities for feedback, reflections, and revision (NRC 1999). Assessment experts strongly recommend that what is tested should reflect what is taught (NCTM 2000). When the focus and form of assessment are different from that of instruction, assessment subverts students' learning ..." (NCTM 1995, p. 13). The evidence suggests that the use of calculators can make a positive difference in what students know and are able to do. Yes, calculators can and should play an important role in assessment, a role mitigated by an understanding of how the technology interacts with the nature of the tasks given and the learning outcomes desired, but a role that can lead to a deeper and more profound understanding of mathematics by more students.

Universal Design and Conserving the Mathematical Construct

Developed in architectural design during the 1980s, the Universal Design concept assured accessibility for all in the planning and construction of buildings, parking lots, and throughways. Replacing step-down curbs with wheelchair accessible ramps is one example of such work. Today, researchers at the University of Minnesota are applying these same principles to the design of student assessment instruments. The goal of Universal Design in Student Assessment is to "...provide access to assessment questions to allow the greatest number of students the opportunity to demonstrate their knowledge and skills" (Thurlow, Quenemoen, Thompson, and Lehr 2001). Early work on Universal Design in Student Assessment focused primarily on linguistic and formatting issues that were common across all content areas. However, just as each type of architectural structure is form- and function-specific, so too is each content area assessment.

To address specific concerns in mathematics, Petit and Lager (2003) have adapted the principles of Universal Design to craft "Conserving the Mathematical Construct" (CMC). These adapted criteria have been used to design and review items for inclusion in this assessment sampler. This chapter outlines the criteria used and references specific items in the sampler.

Criteria of Conserving the Mathematical Construct (CMC)

1. Explicitly align items with the content and cognitive demands in the expectations (standards, objectives, grade-level expectations) being assessed
2. Make intentional decisions about when and why an item is embedded in a context
3. Streamline the language of the item
4. Appropriately use representations (e.g., graphics, pictures, graphs, tables, diagrams, and models) when they are essential to the mathematics being assessed

Applying the criteria of CMC to items in this sampler

Explicitly align items. Each question in this assessment sampler is aligned with the standards and expectations set forth in *Principles and Standards for School Mathematics* (NCTM 2000). In addition, the specific related mathematics being assessed is identified. Each question has been pilot tested, and the students' work has been reviewed.

Intentional decisions about context. Some of the assessment items in this sampler are presented in a context and some are not; these intentional decisions have been based on rationale identified by Treffers and Goffree (cited in DeLange [1984]):

- the context is realistic given mathematics;
- the context is accurate;
- the context serves as a support for student thinking; and
- the context avoids common "pitfalls" (e.g., overuse, contrivance, ambiguity, or bias).

Streamline language. Universal design has focused on researching the best accommodations for English language learners (ELLs), and therefore language has been a major focus. Conserving the Mathematical Construct extends this access to all students. Evidence indicates that some students, for whom English is a first and only language, struggle with language issues, as well (Lager 2002). Anecdotally, teacher observations support this finding.

Although Conserving the Mathematical Construct draws on earlier work[3], it goes beyond these works on language by focusing clearly on the mathematical construct; it deals with mathematical and nonmathematical language and with issues that arise involving language structures.

---

3 See Kopriva (2000) on readability and the amount of text students are expected to use; see Hanson et al. (1998), Kopriva (2000), and Abedi et al. (2001a, 2001b) in regard to Simplified Language.

Over the past decade educators have made efforts to build problems in engaging contexts with interesting, but often complicated, language. Two versions of the problem Cool Jeans Hut are found below to exemplify this point. Both versions are embedded in a context familiar to most students (purchasing jeans), and both include important mathematics. The original version has extraneous language to engage students. The streamlined version conserves the mathematical construct (i.e., compound discounts) and the engagement, but it streamlines the language and modifies the format to highlight important elements of the problem.

Cool Jeans Hut
Mathematical construct

Content Standard (NCTM 2000):	Understand numbers, ways of representing numbers, relationships among numbers, and number systems
Expectation:	Work flexibly with fractions, decimals, and percents to solve problems
Specific mathematics:	Applying compound discounts
Rationale for context:	Discounting merchandise is a common application of percents.

Cool Jeans Hut—Original
(Extraneous language underlined)

Graham loves to shop at Cool Jeans Hut, and received a valued customer coupon in the mail that is good for 30% off any one pair of jeans. When he arrived at the store, he was thrilled to find it was Twenty Tuesday and each customer would get 20% off of his or her purchase. Graham picked out a pair of jeans and went to the checkout. The clerk looked worried, then said, "Oh, 30% and 20% is 50% off, so you get them for $20." At the end of the shift, the manager questioned the clerk about Graham's purchase. The manager said, "You undercharged Graham. I am going to take it out of your pay." How much will the manager take from the clerk's check? Is that the only possibility? Explain the problem involved here. What is the actual discount Graham should have received?

Cool Jeans Hut—Revised

Graham received a coupon in the mail for a 30% discount on any pair of jeans at Cool Jeans Hut. When he arrived at the store, there was an additional 20% discount off all purchases.

Graham picked out a pair of jeans for $40.00. The clerk added the 30% and 20% discounts together and charged Graham $20.

Later, the manager of the store told the clerk that he had undercharged Graham.

A) By how much did the clerk undercharge Graham?
B) Explain to the clerk what he did wrong. What should the clerk do in the future when there is more than one discount on an item?

Use of representations: Representations (graphics, pictures, graphs, tables, diagrams, and models) are essential in mathematics. However, the inclusion of representations in items should be intentional. The representations included should be–

- appropriate,
- mathematically accurate, and
- necessary to the problem

Conclusion

Crucial to instructional decision making is the ability for teachers to obtain quality information about students' understanding of mathematics. Yet, too often in assessment, items written for students' use embody many of the shortcomings that have been raised in this paper. In an effort to engage students, teachers have often over-contextualized assessment items, used superfluous or obtuse language, or included unnecessary or poorly drawn diagrams or pictures that get in the way of obtaining the information necessary to solve a problem. Using the criteria set forth by Conserving the Mathematical Construct can help teachers improve the assessment tasks they create or select for students.

The following steps can help teachers apply the basic principles of Conserving the Mathematical Construct as they refine assessment questions that have been used in the past or as they develop new questions:

1. Identify the mathematical construct being assessed (the mathematics; the expectation, standard, or objective; and the level of complexity).
2. Identify the context or contexts used.
3. Decide which context or contexts, if any, are important to the mathematics being assessed.
4. Provide a rationale for this decision.
5. Identify and repair linguistic inconsistencies, unnecessary language, inconsistent verb tenses, and words that have multiple meanings.
6. Review representations for clarity and importance to the construct being assessed.

For additional information on Universal Design in Assessment and Conserving the Mathematical Construct, we refer the reader to Abedi and Lord (2001); DeLange (1984); Hanson and others (1998); Kopriva 2000; Lager and Petit (2003); and Thurlow, Quenemoen, Thompson, and Lehr (2001).

Appendix
Items Matrices

Number and Operations Items Matrix, Chapter 1

Assessment Item Number	1	2	3	4	5	6	7	8	9	10	11	12	13	14	15	16
Standards and Expectations																
Understand numbers, ways of representing numbers, relationships among numbers, and number systems	X	X	X	X	X	X	X	X	X	X	X	X	X	X	X	X
Work flexibly with fractions, decimals, and percents to solve problems	X	X	X		X	X		X								
Compare and order fractions, decimals, and percents efficiently and find their approximate locations on a number line					X	X										
Develop meaning for percents greater than 100 and less than 1							X	X								
Understand and use ratios and proportions to represent quantitative relationships							X	X	X	X	X					
Develop an understanding of large numbers and recognize and appropriately use exponential, scientific, and calculator notation									X	X	X					
Use factors, multiples, prime factorization, and relatively prime numbers to solve problems												X	X	X		
Develop meaning for integers and represent and compare quantities with them															X	X
Process Standards																
Problem Solving	X	X	X	X	X	X	X	X								X
Communication				X	X	X	X	X	X	X	X				X	
Reasoning and Proof						X	X	X	X	X	X	X	X	X	X	X
Connections								X	X	X	X	X	X	X		
Representation								X	X	X	X	X	X	X		
Item Format	SR	MC	SR	SR	SR	SR	MC	SR	SR	SR	SR	SR	SR	SR	SR	MC

Number and Operations Items Matrix, Chapter 1—*Continued*

Assessment Item Number	1	2	3	4	5	6	7	8	9	10	11	12	13	14	15	16
Standards and Expectations																
Understand meanings of operations and how they relate to one another	X														X	X
Understand the meaning and effects of arithmetic operations with fractions, decimals, and integers															X	X
Use the associative and commutative properties of addition and multiplication and the distributive property of multiplication over addition to simplify computations with integers, fractions, and decimals	X															
Compute fluently and make reasonable estimates					X	X		X			X				X	X
Select appropriate methods and tools for computing with fractions and decimals from among mental computation, estimation, calculators or computers, and paper and pencil, depending on the situation, and apply the selected methods					X	X					X				X	X
Develop and analyze algorithms for computing with fractions, decimals, and integers and develop fluency in their use								X								
Develop, analyze, and explain methods for solving problems involving proportions, such as scaling and finding equivalent ratios							X									
Process Standards																
Problem Solving	X	X	X	X	X	X	X	X	X	X	X					X
Communication				X	X	X	X	X	X	X	X				X	
Reasoning and Proof						X	X	X	X	X	X	X	X	X	X	X
Connections								X	X	X		X	X			
Representation								X	X	X	X	X	X	X		
Item Format	SR	MC	SR	SR	SR	SR	MC	SR	SR	SR	SR	SR	SR	SR	SR	MC

Number and Operations Items Matrix, Chapter 1—Continued

Assessment Item Number	17	18	19	20	21	22	23	24	25	26	27	28	29	30	31	32	33
Standards and Expectations																	
Understand numbers, ways of representing numbers, relationships among numbers, and number systems	X	X	X	X	X	X	X	X					X				
Work flexibly with fractions, decimals, and percents to solve problems	X	X					X	X					X				
Develop meaning for integers and represent and compare quantities with them	X	X	X	X	X	X			X			X					
Understand meanings of operations and how they relate to one another	X	X	X	X	X	X	X	X				X					
Understand the meaning and effects of arithmetic operations with fractions, decimals, and integers			X	X	X		X	X				X					
Use the associative and commutative properties of addition and multiplication and the distributive property of multiplication over addition to simplify computations with integers, fractions, and decimals					X	X			X								
Understand and use the inverse relationships of addition and subtraction, multiplication and division, and squaring and finding square roots to simplify computations and solve problems							X		X								
Process Standards																	
Problem Solving	X	X	X	X	X	X	X	X	X	X	X	X	X		X	X	
Communication									X	X							
Reasoning and Proof	X	X	X	X	X	X	X		X		X	X	X	X	X	X	
Connections			X	X													
Representation		X			X			X	X								
Item Format	MC	MC	SR	SR	MC	SR	SR	SR	ER	SR	MC	SR	SR	SR	SR	MC	

Number and Operations Items Matrix, Chapter 1—*Continued*

Assessment Item Number	17	18	19	20	21	22	23	24	25	26	27	28	29	30	31	32	33
Standards and Expectations																	
Compute fluently and make reasonable estimates							X	X	X	X	X	X	X	X	X	X	X
Select appropriate methods and tools for computing with fractions and decimals from among mental computation, estimation, calculators or computers, and paper and pencil, depending on the situation, and apply the selected methods							X	X	X	X	X	X					
Develop and analyze algorithms for computing with fractions, decimals, and integers and develop fluency in their use								X	X				X	X			
Develop and use strategies to estimate the results of rational-number computations and judge the reasonableness of the results															X	X	
Develop, analyze, and explain methods																	X
Process Standards																	
Problem Solving		X	X	X			X	X	X	X	X	X	X	X	X	X	X
Communication									X	X							
Reasoning and Proof	X	X	X	X	X	X	X		X		X	X	X	X	X	X	X
Connections		X	X	X													
Representation		X			X			X	X								
Item Format	MC	MC	SR	SR	MC	SR	SR	SR	ER	SR	MC	SR	SR	SR	SR	MC	MC

Algebra Items Matrix, Chapter 2

Assessment Item Number	1	2	3	4	5	6	7	8	9	10	11	12
Standards and Expectations												
Understand patterns, relations, and functions	X	X	X	X	X	X						
Represent, analyze, and generalize a variety of patterns with tables, graphs, words, and, when possible, symbolic rules	X	X	X	X		X						
Relate and compare different forms of representation for a relationship				X	X							
Identify functions as linear or nonlinear and contrast their properties from tables, graphs, or equations						X						
Represent and analyze mathematical situations and structures using algebraic symbols							X	X	X	X	X	
Develop an initial conceptual understanding of different uses of variables							X					
Explore relationships between symbolic expressions and graphs of lines, paying particular attention to the meaning of intercept and slope								X	X			
Use symbolic algebra to represent situations and to solve problems, especially those that involve linear relationships									X			
Recognize and generate equivalent forms for simple algebraic expressions and solve linear equations										X	X	
Use mathematical models to represent and understand quantitative relationships												X
Model and solve contextualized problems using various representations, such as graphs, tables, and equations												X
Process Standards												
Problem Solving						X	X			X	X	
Communication		X				X	X		X			
Reasoning and Proof	X	X	X	X	X			X			X	
Connections									X	X		
Representation				X				X			X	X
Item Format	SR	SR	SR	SR	SR	SR	MC	MC	SR	SR	SR	MC

Algebra Items Matrix, Chapter 2—Continued

Assessment Item Number	13	14	15	16	17	18	19	20	21	22
Standards and Expectations										
Represent and analyze mathematical situations and structures using algebraic symbols							X	X		
Explore relationships between symbolic expressions and graphs of lines, paying particular attention to the meaning of intercept and slope							X	X		
Use mathematical models to represent and understand quantitative relationships	X	X	X	X	X	X				
Model and solve contextualized problems using various representations, such as graphs, tables, and equations	X	X	X	X	X	X				
Analyze change in various contexts				X	X	X	X	X	X	X
Use graphs to analyze the nature of changes in quantities in linear relationships				X	X	X	X	X	X	X
Process Standards										
Problem Solving	X			X	X	X	X	X		
Communication						X		X		
Reasoning and Proof	X	X	X	X	X	X	X	X	X	X
Connections				X	X	X			X	
Representation		X	X		X	X	X	X	X	X
Item Format	SR	MC	MC	ER	ER	ER	SR	SR	SR	SR

Geometry Items Matrix, Chapter 3

Assessment Item Number	1	2	3	4	5	6	7	8	9	10	11	12	13	14
Standards and Expectations														
Analyze characteristics and properties of two- and three- dimensional shapes and develop mathematical arguments about geometric relationships	X	X	X	X	X	X	X	X	X	X	X			
Precisely describe, classify, and understand relationships among types of two- and three-dimensional objects using their defining properties	X	X	X	X	X									
Understand relationships among the angles, side lengths, perimeters, areas, and volumes of similar objects						X	X	X	X					
Create and critique inductive and deductive arguments concerning geometric ideas and relationships, such as congruence, similarity, and the Pythagorean relationship										X	X			
Specify locations and describe spatial relationships using coordinate geometry and other representational systems												X	X	
Use coordinate geometry to represent and examine the properties of geometric shapes												X	X	
Use coordinate geometry to examine special geometric shapes, such as regular polygons or those with pairs of parallel or perpendicular sides												X	X	
Apply transformations and use symmetry to analyze mathematical situations														X
Describe sizes, positions, and orientations of shapes under informal transformations such as flips, turns, slides, and scaling														X
Process Standards														
Problem Solving	X		X	X	X	X	X	X	X	X	X			X
Communication		X		X			X		X	X				
Reasoning and Proof					X			X				X	X	X
Connections											X	X		
Representation									X		X			
Item Format	MC	SR	MC	SR	SR	MC	SR	MC	ER	SR	SR	MC	SR	MC

Geometry Items Matrix, Chapter 3—Continued

Assessment Item Number	15	16	17	18	19	20	21	22	23	24	25	26	27
Standards and Expectations													
Analyze characteristics and properties of two- and three- dimensional shapes and develop mathematical arguments about geometric relationships								X					
Understand relationships among the angles, side lengths, perimeters, areas, and volumes of similar objects									X				
Specify locations and describe spatial relationships using coordinate geometry and other representational systems	X	X											
Use coordinate geometry to examine special geometric shapes, such as regular polygons or those with pairs of parallel or perpendicular sides	X	X											
Apply transformations and use symmetry to analyze mathematical situations	X	X	X	X	X	X							
Describe sizes, positions, and orientations of shapes under informal transformations such as flips, turns, slides, and scaling	X	X	X										
Examine congruence, similarity, and line or rotational symmetry of objects using transformations				X	X	X							
Use visualization, spatial reasoning, and geometric modeling to solve problems							X		X	X	X	X	X
Draw geometric objects with specific properties, such as side lengths or angle measures							X			X			
Use two-dimensional representations of three-dimensional objects to visualize and solve problems such as those involving surface area and volume									X	X	X		
Use geometric models to represent and explain numerical and algebraic relationships										X			
Recognize and apply geometric ideas and relationships in areas outside the mathematics classroom, such as art, science, and everyday life												X	X
Process Standards													
Problem Solving	X	X				X	X	X	X	X	X	X	X
Communication							X	X					
Reasoning and Proof	X	X	X	X	X			X	X		X	X	X
Connections													
Representation						X				X			
Item Format	MC	SR	MC	MC	MC	SR	SR	ER	MC	SR	SR	SR	SR

Measurement Items Matrix, Chapter 4

Assessment Item Number	1	2	3	4	5	6	7	8	9	10	11	12	13	14	15
Standards and Expectations															
Understand measurable attributes of objects and the units, systems, and processes of measurement	X	X	X	X	X	X	X	X	X	X	X	X	X	X	X
Understand both metric and customary systems of measurement	X	X	X	X	X	X	X	X	X	X	X	X	X	X	X
Understand relationships among units and convert from one unit to another within the same system	X	X	X	X	X			X							
Understand, select, and use units of appropriate size and type to measure angles, perimeter, area, and volume			X		X	X	X	X							
Apply appropriate techniques, tools, and formulas to determine measurements		X	X	X	X	X	X	X	X	X	X	X	X	X	X
Use common benchmarks to select appropriate methods for estimating measurements				X				X	X						
Select and apply techniques and tools to accurately find length, area, volume, and angle measures to appropriate levels of precision			X			X			X	X	X	X	X	X	X
Develop and use formulas to determine the circumference of circles and the area of triangles, parallelograms, trapezoids, and circles and develop strategies to find the area of more-complex shapes						X	X	X							
Develop strategies to determine the surface area and volume of selected prisms, pyramids, and cylinders						X									
Solve problems involving scale factors, using ratio and proportion							X	X							
Solve simple problems involving rates and derived measurements for such attributes as velocity and density															
Process Standards															
Problem Solving	X		X	X	X	X	X	X	X	X	X	X	X	X	X
Communication		X	X						X		X			X	X
Reasoning and Proof			X	X	X	X		X	X	X	X	X	X	X	X
Connections							X		X				X		
Representation					X		X	X							
Item Format	SR	SR	SR	SR	SR	SR	ER	ER	SR	SR	SR	MC	MC	MC	SR

Measurement Items Matrix, Chapter 4—Continued

Assessment Item Number	16	17	18	19	20	21	22	23	24	25	26	27	28	29
Standards and Expectations														
Understand measurable attributes of objects and the units, systems, and processes of measurement	X	X	X	X	X	X	X	X	X	X	X	X	X	X
Understand both metric and customary systems of measurement	X	X	X	X	X	X	X	X	X	X	X	X	X	X
Understand, select, and use units of appropriate size and type to measure angles, perimeter, area, surface area, and volume													X	X
Apply appropriate techniques, tools, and formulas to determine measurements	X	X	X	X	X	X	X	X	X	X	X	X	X	X
Select and apply techniques and tools to accurately find length, area, volume, and angle measures to appropriate levels of precision	X									X				
Develop and use formulas to determine the circumference of circles and the area of triangles, parallelograms, trapezoids, and circles and develop strategies to find the area of more-complex shapes		X	X	X										
Develop strategies to determine the surface area and volume of selected prisms, pyramids, and cylinders					X	X	X	X	X					
Solve problems involving scale factors, using ratio and proportion											X	X	X	X
Solve simple problems involving rates and derived measurements for such attributes as velocity and density										X				
Process Standards														
Problem Solving	X	X	X	X	X			X	X	X	X	X	X	X
Communication	X									X				
Reasoning and Proof	X				X	X	X	X	X	X			X	X
Connections														
Representation														
Item Format	SR	SR	SR	MC	SR	MC	MC	SR	ER	SR	SR	MC	SR	SR

Data Analysis and Probability Items Matrix, Chapter 5

Assessment Item Number	1	2	3	4	5	6	7	8	9	10	11	12	13
Standards and Expectations													
Formulate questions that can be addressed with data and collect, organize, and display relevant data to answer them	X												
Formulate questions, design studies, and collect data about a characteristic shared by two populations or different characteristics within one population	X												
Select, create, and use appropriate graphical representations of data, including histograms, box plots, and scatterplots	X												
Select the appropriate statistical methods to analyze data	X	X	X	X	X	X	X	X	X	X	X	X	X
Find, use, and interpret measures of center and spread, including mean and interquartile range		X	X	X	X	X	X	X	X	X			
Discuss and understand the correspondence between data sets and their graphical representations, especially histograms, stem-and-leaf plots, box plots, and scatterplots	X										X	X	X
Develop and evaluate inferences and predictions that are based on data													X
Use observations about differences between two or more samples to make conjectures about the populations from which the samples were taken													X
Process Standards													
Problem Solving	X									X			X
Communication			X		X				X		X		
Reasoning and Proof	X	X		X	X	X	X	X	X	X	X	X	X
Connections				X									
Representation	X	X	X			X		X		X		X	X
Item Format	ER	MC	MC	MC	SR	MC	MC	MC	SR	ER	SR	SR	ER

Data Analysis and Probability Items Matrix, Chapter 5—Continued

Assessment Item Number	14	15	16	17	18	19	20	21	22
Standards and Expectations									
Develop and evaluate inferences and predictions that are based on data	X	X	X	X	X				
Use observations about differences between two or more samples to make conjectures about the populations from which the samples were taken		X	X	X					
Make conjectures about possible relationships between two characteristics of a sample on the basis of scatterplots of the data and approximate lines of fit		X							
Use conjectures to formulate new questions and plan new studies to answer them	X		X		X				
Understand and apply basic concepts of probability				X	X	X	X	X	X
Understand and use appropriate terminology to describe complementary and mutually exclusive events				X		X			
Use proportionality and a basic understanding of probability to make and test conjectures about the results of experiments and simulations					X	X	X		
Compute probabilities from simple compound events, using such methods as organized lists, tree diagrams, and area models							X	X	X
Process Standards									
Problem Solving	X	X		X					X
Communication	X		X	X	X	X	X		
Reasoning and Proof	X	X	X	X	X	X	X	X	X
Connections									
Representation	X		X						
Item Format	SR	SR	SR	SR	MC	SR	SR	SR	SR

Bibliography

Abedi, Jamal, and Carol Lord. "The Language Factor in Mathematics Tests." *Applied Measurement in Education* 14 (3) (2001): 219–34.

American Educational Research Association. "AERA Position Statement Concerning High-Stakes Testing in PreK–12 Education." 2000. http://www.aera.net/about/policy/stakes.htm (accessed March 12, 2005).

Braswell, James S., Anthony D. Lutkus, Wendy S. Grigg, Shari L. Santapau, Brenda Tay-Lim, and Matthew Johnson. *The Nation's Report Card: Mathematics 2000.* Washington, D.C.: National Center for Educational Statistics. 2001. http://nces.ed.gov/nationsreportcard/pubs/main2000/2001517.asp#section5 (accessed March 12, 2005).

Burrill, Gail. "A Revolution in My High School Classroom." *Mathematics Education Dialogues.* A publication of the National Council of Teachers of Mathematics. May/June 1999, p. 13.

Burrill, Gail F., Jacquie Allison, Glenda Breaux, Signe Kastberg, Keith Leatham, and Wendy Sanchez. *Handheld Graphing Technology in Secondary Schools: Research Findings and Implications for Classroom Practice.* Dallas, Tex.: Texas Instruments, 2003.

Bush, William S., and Steve Leinwand. *Mathematics Assessment: A Practical Handbook for Grades 6–8.* Reston, Va.: National Council of Teachers of Mathematics, 2000.

Carter, Betsy Y., and Steven J. Leinwand. "Calculators and Connecticut's Eighth-Grade Mastery Test." *Arithmetic Teacher* 34 (February 1987): 55–56.

Clements, Douglas, H. "Computers in Mathematics Education Assessment." In *Classroom Assessment in Mathematics,* edited by George W. Bright and Jeane M. Joyner, pp. 153–58. New York: University Press of American, 1998.

DeLange, Jan. *Mathematics Insights and Meaning.* Netherlands: University of Utrecht, 1987.

Fey, James T. "Technology and Mathematics Education: A Survey of Recent Developments and Important Problems." *Educational Studies in Mathematics* 20 (1989): 237–72.

Hanson, Matthew R., John R. Hayes, Kenneth Schriver, Paul LeMahieu, and Pamela J. Brown. *A Plain Language Approach to the Revision of Test Items.* San Diego, Calif.: American Educational Research Association, 1998.

Heid, M. Kathleen. "Soundoff: Calculators on Tests—One Giant Step for Mathematics Education." *Mathematics Teacher* 81 (December 1988): 710–13.

Hembree, Ray, and Donald J. Dessart. "Effects of Hand-held Calculators in Precollege Mathematics Education: A Meta-Analysis." *Journal for Research in Mathematics Education* 17 (2) (March 1986): 83–99.

———. "Research on Calculators in Mathematics Education." In *Calculators in Mathematics Education,* 1992 Yearbook, edited by James T. Fey and Christian R. Hirsch, pp. 23–32. Reston, Va.: National Council of Teachers of Mathematics, 1992.

Hess, Frederick M. "Refining or Retreating? High-Stakes Accountability in the States." In *No Child Left Behind: The Politics and Practice of School Accountability*, edited by Paul E. Peterson and Martin R. West, pp. 55–79. Washington, D.C.: Brookings Institution Press, 2003.

Hopkins, Mary H. "The Use of Calculators in the Assessment of Mathematics Achievement." In *Calculators in Mathematics Education*, 1992 Yearbook, edited by James T. Fey and Christian R. Hirsch, pp. 158–65. Reston, Va.: National Council of Teachers of Mathematics, 1992.

Hudson, Susan B., Kelly C. McMahon, and Christina M. Overstreet. *The 2000 National Survey of Science and Mathematics Educators: Compendium of Tables.* Chapel Hill, N.C.: Horizon Research, 2002.

Kemp, Marian, Barry Kissane, and Jen Bradley. "Graphics Calculator Use in Examinations: Accident or Design?" *Australian Senior Mathematics Journal* 10 (1) (1996): 35–50.

Kenelly, John W., ed. *The Use of Calculators in Standardized Testing of Mathematics.* New York: College Entrance Examination Board, 1989.

Kissane, Barry. "New Calculator Technologies and Examinations." In *Proceedings of the Fifth Asian Technology Conference in Mathematics,* edited by Wu-Chun Yang, Suling Chu, and Jen-Chung Chuan, pp. 365–74. Chang Mai, Thailand: 2000. (ISBN 974-657-362-4). Available at http://wwwstaff.murdoch.edu.au/~kissane (accessed April 19, 2004).

Kopriva, Rebecca. *Ensuring Accuracy in Testing for English Language Learners: A Practical Guide for Assessment.* Washington, D.C.: Council of Chief State School Officers, SCASS-LEP Consortium, 2000.

Lager, Carl A. "Language and Mathematics: Improving Algebra Instruction for English Learners." Unpublished dissertation, 2002.

Lager, Carl, and Marge Petit. "Conserving the Mathematical Construct." TSNE Mathematics Test Specifications (Version 12.0). October 15, 2003. www.nciea.org (accessed March 22, 2005).

Leinwand, Steven J. "Calculators in State Testing: A Case Study." In *Calculators in Mathematics Education,* 1992 Yearbook, edited by James T. Fey and Christian R. Hirsch, pp.167–76. Reston, Va.: National Council of Teachers of Mathematics, 1992.

Long, Vena M., Barbara Reys, and Steven J. Osterlind. "Using Calculators on Achievement Tests." *Mathematics Teacher* 82 (May 1989): 318–25.

Lutzer, David J., James W. Maxwell, and Stephen B. Rodi. *Statistical Abstract of Undergraduate Programs in the Mathematical Sciences in the United States: Fall 2000 CBMS Study.* Washington, D.C.: Mathematical Association of America, 2002.

National Assessment Governing Board. *Mathematics Framework for the 1996 National Assessment of Educational Progress.* NAEP Mathematics Consensus Project. Washington D.C.: U.S. Government Printing Office, 1996.

National Assessment of Educational Progress (NAEP). *2005 Mathematics NAEP Framework.* Washington, D.C.: Council of Chief State School Officers, 2001.

National Center for Educational Statistics. "Long Term Trend Results in Mathematics." 2000. http://nces.ed.gov/nationsreportcard/mathematics/trends.asp (accessed March 22, 2005).

———. "NNAEP Mathematics—Student Report on Frequence of Calculator Use." 2000. http://nces.ed.gov/nationsreportcard/mathematics/results/calculator.asp (accessed March 22, 2005).

National Council of Teachers of Mathematics (NCTM). *Assessment Standards for School Mathematics.* Reston, Va.: NCTM, 1995.

———. *Curriculum and Evaluation Standards for School Mathematics.* Reston Va.: NCTM, 1989.

———. *Principles and Standards for School Mathematics.* Reston Va.: NCTM, 2000.

———. *Professional Standards for Teaching Mathematics.* Reston, Va.: NCTM, 1991.

National Research Council. *Adding It Up: Helping Children Learn Mathematics,* edited by Jeremy Kilpatrick, Jane Swafford, and Bradford Findell. Washington, D.C.: National Academy Press, 2001.

————. *How People Learn: Brain, Mind, Experience, and School*, edited by John D. Bransford, Ann L. Brown, and Rodney R. Cocking. Washington, D.C.: National Academy Press, 1999.

————. *Measuring What Counts*. Washington D.C.: National Academy Press, Mathematical Sciences Education Board, 1993.

North Carolina State Department of Public Instruction. *Mathematics Curriculum Study: A Report of the Mathematics Curriculum Study Committee to the North Carolina State Board of Education*. Raleigh, N.C.: North Carolina State Department of Public Instruction, 1983.

Oregon State Department of Education. *Mathematics Grade 8: Oregon Statewide Assessment, 1985*. Salem, Ore.: Oregon State Department of Education, 1985.

Shannon, Ann. *Keeping Score*. Washington, D.C.: National Academy Press, Center for Science, Mathematics, and Engineering Education, 1999.

Smith, Brian A. "A Meta-Analysis of Outcomes from the Use of Calculators in Mathematics Education." *Dissertation Abstracts International* 58 (1997): 787A.

Stenmark, Jean, ed. *Mathematics Assessment: Myths, Models, Good Questions, and Practical Suggestions*. Reston, Va.: National Council of Teachers of Mathematics, 1991.

Thirteen Ed Online. "Concept to Classroom: A Series of Workshops." 2004. http://www.thirteen.org/edonline/concept2class/standards/explor_sub2.html (accessed March 12, 2005).

Thurlow, Martha, Rachel Quenemoen, Sandra Thompson, and Camilla Lehr. "Principles and Characteristics of Inclusive Assessment and Accountability Systems." Synthesis Report 40. Minneapolis, Minn: National Center on Educational Outcomes. Available at http://education.umn.edu/nceo/OnlinePubs/Synthesis40.html (accessed March 22, 2005).

Tolbert, Clara. "Evidence of Effective Instruction Using Handheld Technology." Presentation at the Annual Meeting of the National Council of Supervisors of Mathematics, Philadelphia, Pa., 2004.

Usiskin, Zalman. "Groping and Hoping for a Consensus on Calculator Use." *Mathematics Education Dialogues*. A publication of the National Council of Teachers of Mathematics. May/June 1999a, p. 1.

Usiskin, Zalman, ed. *Mathematics Education Dialogues*. A publication of the National Council of Teachers of Mathematics. Theme Issue: Calculators. May/June 1999b.

Van den Heuvel-Panhuizen, Maria. *Assessment and Realistic Mathematics Education.* Utrecht, Netherlands: Freudenthal Institute, 1996.

Webb, Norman. *Criteria for Alignment of Expectations and Assessments on Mathematics and Science Education.* Research Monograph Number 6. Washington, D.C.: Council of Chief State School Officers, 1997.

————. "Depth of Knowledge Levels for Four Content Areas." March 2002. www .education.ky.gov/.../ DOKHandOutI.doc (accessed March 22, 2005).

Wilson, James W., and Jeremy Kilpatrick. "Theoretical Issues in Development of Calculator-based Mathematics Tests." In *The Use of Calculators in Standardized Testing of Mathematics,* edited by John W. Kenelly. New York: College Entrance Examination Board, 1989.

Sources for Assessment Items

ACT. "WorkKeys Assessments, Applied Mathematics." 2003. http://www.act.org/workkeys/assess/math/sample7.html (accessed March 22, 2005).

Applied Mathematics. Emeryville, Calif.: Key Curriculum Press, 1999. Contact Key Curriculum Press, 1150 65th Street, Emeryville, CA 94608; 1-800-995-MATH; www.keypress.com.

The Chiswick Centre (414 Chiswick High Road, London, W4 5TF), copyright © QCA 2003. http://www.worldclassarena.org/v5/example_questions.htm (accessed March 22, 2005).

The Connected Mathematics Project. Upper Saddle River, N.J.: Prentice Hall, 2002.

Connecticut State Department of Education.http://www.state.ct.us/sde/dtl/curriculum/ CAPT2g/capt_43-50.pdf item 5 (accessed March 22, 2005).

Gnanadesikian, Mrudulla. *The Art and Techniques of Simulation.* Palo Alto, Calif.: Dale Seymour Publications, 1987.

Maine Department of Education. "Maine 2002 Educational Assessment, Grade 8, Released Items." http://mainegovimages.informe.org/education/mea/2002ReleasedItems/ Mar02G4MathScoreResponses.pdf (accessed March 22, 2005).

MARS and Balanced Assessment Resource Service. http://www.educ.msu.edu/MARS (accessed March 22, 2005).

Massachusetts Department of Education. "The Massachusetts Comprehensive Assessment System, Release of Spring 2001 Eighth-Grade Mathematics Items." http://www.doe.mass.edu/mcas/2001/release/ (accessed March 22, 2005).

Mathematics Achievement Partnership. *Foundations of Success: Mathematics for the Middle Grades*. Washington, D.C.: Achieve, 2001.

Mathematics in Context. New York: Holt, Rinehart &Winston, © 2000 by Encyclopædia Britannica.

The Math Forum @ Drexel. http://mathforum.org/midpow/solutions/19980511.midpow .html (accessed March 22, 2005).

MathThematics, Vol. 2. New York: Houghton Mifflin Co., © 1998 by McDougal Littell, Evanston, Illinois.

MathThematics, Vol. 3. New York: Houghton Mifflin Co., © 1998 by McDougal Littell, Evanston, Illinois.

New England Common Assessment Program Grade Level Expectations. "Mathematics GLE Resource Materials. http://64.233.161.104/search?q=cache:bSlrfK4boI4J: www.state.vt.us/educ/new/pdfdoc/pubs/grade_expectations/math_resources/ data_statistics.pdf++NECAP+GLE&hl=en (accessed April 18, 2005).

Nova Scotia Department of Education. *Nova Scotia Junior High Mathematics Program Assessment*. Halifax: Nova Scotia Department of Education, 2003, 2004. http://plans.ednet.ns.ca/

Performance Tasks and Rubrics: Middle School Mathematics. Larchmont, N.Y.: Eye on Education, 1997.

Shapes and Space. MathScapes Program. Columbus, Ohio: Glencoe/McGraw-Hill, copyright 1998 by Creative Publications, DeSoto Texas.

St. Francis Xavier University, Nova Scotia. http://www.stfx.ca/special/mathproblems (accessed March 22, 2005).

Virginia Department of Education. "Virginia Standards of Learning Assessments, Eighth-Grade Mathematics, Spring 2002 Released Tests." http://www.pen.k12.va.us/VDOE/ Assessment/Release2002/Grade8/VirgOnLine_Gr8_Math_1-23.pdf Item #58 (accessed March 22, 2005).

Three additional titles
are planned for the
Mathematics Assessment Samplers series

Anne M. Collins, series editor

○ *Mathematics Assessment Sampler, Prekindergarten–Grade 2: Items Aligned with NCTM's* Principles and Standards for School Mathematics, edited by DeAnn Huinker

---

○ *Mathematics Assessment Sampler, Grades 3–5, Items Aligned with NCTM's* Principles and Standards for School Mathematics, edited by Jane D. Gawronski

---

○ *Mathematics Assessment Sampler, Grades 9–12, Items Aligned with NCTM's* Principles and Standards for School Mathematics, edited by Betty Travis

---

Please consult www.nctm.org/catalog for the availability of these titles, as well as for a plethora of resources for teachers of mathematics at all grade levels.

For the most up-to-date listing of NCTM resources on topics of interest to mathematics educators, as well as information on membership benefits, conferences, and workshops, visit the NCTM Web site at www.nctm.org.